AF818632

SOTHIS PRESS

Sothis Press aims to completely reimagine the public domain book with an emphasis on universal access to captivatingly designed print editions. We believe that technology will enable every single book ever published to be perpetually accessible in beautiful print form. Sothis is beauty in ubiquity.

www.sothis.press
1 Waterhouse Square, London, EC1N 2ST

A Manual of Field Astronomy

Andrew H. Holt

SOTHIS PRESS

ISBN 978-1-0365-1039-8

A Manual of Field Astronomy
is in the public domain.

This edition is published by
Sothis Press © 2023. The company
is registered in England and Wales.
The Company Number is 14900389.

A MANUAL OF FIELD ASTRONOMY

BY

ANDREW H. HOLT

Instructor in Civil Engineering in the College of Applied Science
of the State University of Iowa

FIRST EDITION
FIRST THOUSAND

NEW YORK
JOHN WILEY & SONS, Inc.
London: CHAPMAN & HALL, Limited
1917

PUBLISHERS PRINTING COMPANY
207-217 West Twenty-fifth Street, New York

PREFACE

If the reason be demanded for the appearance of another book on Field Astronomy when there are already published several excellent works on the subject, it may be stated as follows: That although any one of them may serve very well as a text for a comparatively extended study, the author has been unable to find one sufficiently concise to fit the short time usually allowed for the work in a civil engineering course which would still provide enough of the fundamentals of the subject to enable the reader to make, intelligently, the observations and accompanying computations required in the practice of general engineering and surveying. Something is needed more complete than the usual chapter in books on surveying and less extensive than most texts on field astronomy. This need, which is acknowledged by other teachers to exist, it is hoped to fill; and at the same time it has been attempted to provide a book which will be of service to engineers and surveyors whose practice requires that they occasionally make astronomical observations.

To this end the discussion of fundamentals has been made brief, but it is thought sufficiently thorough for the purpose. Special attention has been given to the matter of measurement of time, because it is believed that this causes more difficulty for students in general than any other part of the subject.

In the selection of the methods described for the determination of latitude, azimuth, time, and longitude, care has been taken to choose those which are believed to be most capable of producing results when used with field instruments under ordinary field conditions. Realizing that the determination of azimuth is more frequently required than any other observation, more methods have been given for this than for the other problems.

Each observation has been presented essentially as follows: The work of which the observation consists is first stated briefly, followed by the relations and theory on which it depends, accompanied by such explanation as seems necessary. The procedure is then outlined, step by step, under the general headings:

"Computations Preceding Field Work," "Field Work," and "Computations Following Field Work." This outline is supplemented by a copy (near the back of the book) of the field-notes and computations of a similar observation.

It is hoped that this method of presentation will commend itself not only to the student but to the engineer in practice.

The "Summary of Observations" in Chapter *XI* should be useful in selecting an observation or in determining whether sufficient data are at hand to permit an observation which is under consideration.

In Appendix A are given the derivations of the formulas of Spherical Trigonometry which are needed in the work, and in Appendix B is a brief discussion of the theory and use of the "Solar Attachment" for the engineer's transit.

No excuse is made for the omission of refinements of either theory or practice which are not required in work done with an engineer's transit or a sextant.

While preparing the manuscript the author has studied several of the existing works on field astronomy, and this book has profited thereby, acknowledgment being made in the body of the book whenever anything has been copied. No claim is made to having produced anything new; but merely to having put well-known facts in a new, and it is hoped useful, form.

The thanks of the author are due to Messrs. W. and L. E. Gurley and the Bausch & Lomb Optical Company, who have furnished cuts for the book; to the Superintendent of the U. S. Coast and Geodetic Survey, who has permitted the use of tables from Government publications, to friends who have given advice and suggestions, and among these particularly to Mr. R. B. Kittredge, Assistant Professor of Railroad Engineering in the College of Applied Science of the State University of Iowa, who has read the entire manuscript, very much to its improvement.

A. H. Holt.

Iowa City, Iowa, November, 1916.

NOTATION

ϕ = Latitude.
λ = Longitude.
Z_n = Azimuth, referred to true north.
Z_s = Azimuth, referred to true south.
t = Hour angle.
RA = Right ascension.
h = Altitude.
δ = Declination.
Z = Interior angle of the astronomical triangle at the zenith.
P = Interior angle of the astronomical triangle at the pole.
S = Interior angle of the astronomical triangle at the star.
z = Co-declination or polar distance, $90° - \delta$.
p = Co-altitude or zenith distance, $90° - h$.
s = Co-latitude, $90° - \phi$.
$k = \frac{1}{2}(s + p + z) = \frac{1}{2}[270° - (\phi + h + \delta)]$.
$Sid.\ T$ = Sidereal time.
$Std.\ T$ = Standard time.
LMT = Local mean time.
LAT = Local apparent time.
A = Right ascension of the mean sun.
A_n = Right ascension of the mean sun at local mean noon.
E = Equation of time.

CONTENTS

CHAPTER IV

THE AMERICAN EPHEMERIS AND NAUTICAL ALMANAC

CHAPTER V

PROBLEMS IN CONVERSION OF TIME

CHAPTER VI

OBSERVATIONS—CORRECTIONS TO OBSERVATIONS

CHAPTER VII

OBSERVATIONS FOR LATITUDE

CHAPTER VIII

OBSERVATIONS FOR AZIMUTH

CHAPTER IX

OBSERVATIONS FOR TIME

CHAPTER X

OBSERVATIONS FOR LONGITUDE

CHAPTER XI

SUMMARY OF OBSERVATIONS

APPENDIX A

SPHERICAL TRIGONOMETRY

APPENDIX B

SOLAR ATTACHMENTS FOR TRANSITS

TABLES

·A MANUAL OF FIELD ASTRONOMY

CHAPTER I

INTRODUCTORY

1. Field Astronomy. Practical Field Astronomy for the engineer consists of the theory and practice of the determination by observations on the sun and the stars of: (1) Latitude, (2) Longitude, (3) Azimuth, and (4) Time. Occasionally observations are made on the moon, but those on the sun and the stars are the most important.

The engineer is not concerned with much that goes to make up the science of Astronomy. He makes measurements of the directions of the heavenly bodies; but takes no account of their distances, their actual movements in space or their physical characteristics. They are to him simply objects of known positions from which he can make measurements to determine his position on the earth's surface or to orient the courses of his survey.

2. The Celestial Sphere. Since only the directions of the sun and stars are to be considered, it is convenient to regard them all as being situated on the surface of a sphere of infinite radius, called the celestial sphere, with its center at the center of the earth. For most of the work the earth (having a finite radius as compared with the infinite radius of the celestial sphere) is considered to be a point. That there is no appreciable error in this is apparent from the fact that the ratio of the radius of the earth to the distance to the nearest fixed star is about 1 to 7,000,000,000.

It should be noted that since the radius of the celestial sphere is assumed to be infinite, all parallel planes which are separated by any finite distance may without appreciable error be considered to cut its surface in the same circle; and all parallel lines may be assumed to pierce it in the same point. Also, any plane through the earth will cut the surface of the celestial sphere in a great circle. (Since it is assumed to pass through the center of the sphere.)

3. Apparent Motion of the Heavenly **Bodies.** Due to the daily rotation of the earth about its axis, all of the stars and the sun appear to be traveling from the east toward the west along circles on the surface of the celestial sphere, making one revolution a day. The earth's axis, produced, would pass through the centers of these circles and would be perpendicular to their planes. Due to the earth's eastward motion in its orbit around the sun, once a year, we see the sun at different times from different places in the orbit, and therefore in apparently different positions. This apparent motion of the sun is eastward along a great circle on the celestial sphere whose plane passes through the earth and makes an angle of about 23° 27′ with the plane of the equator; and it amounts to one revolution around the earth in one year.

In astronomy the terms "east" and "west" are often used to indicate directions of rotation instead of in the sense with which we are familiar in plane surveying. The reason for this will be apparent if one considers two persons standing on opposite sides of the earth and both pointing east (or west). They would actually be pointing in opposite directions, but they would be indicating the same direction of rotation.

The student should become accustomed to thinking of relative positions and motions of objects on the celestial sphere from both an inside and an outside point of view. It is usually easier to visualize these things if one imagines himself outside, with a general view of the whole situation; although, of course, they will actually have to be viewed from an inside position.

If one imagines himself outside the celestial sphere and directly above the north pole, on line with the axis of rotation of the earth, an eastward rotation of a celestial object will appear to be counter-clockwise, while a westward rotation will appear to be clockwise.

In general, we shall study apparent and not real motions; and therefore, whether considering one's self at the center of the sphere or outside and above the north pole, the earth is usually assumed to be standing still and the other bodies to be moving around it.

4. Definitions. The following are some of the terms used in

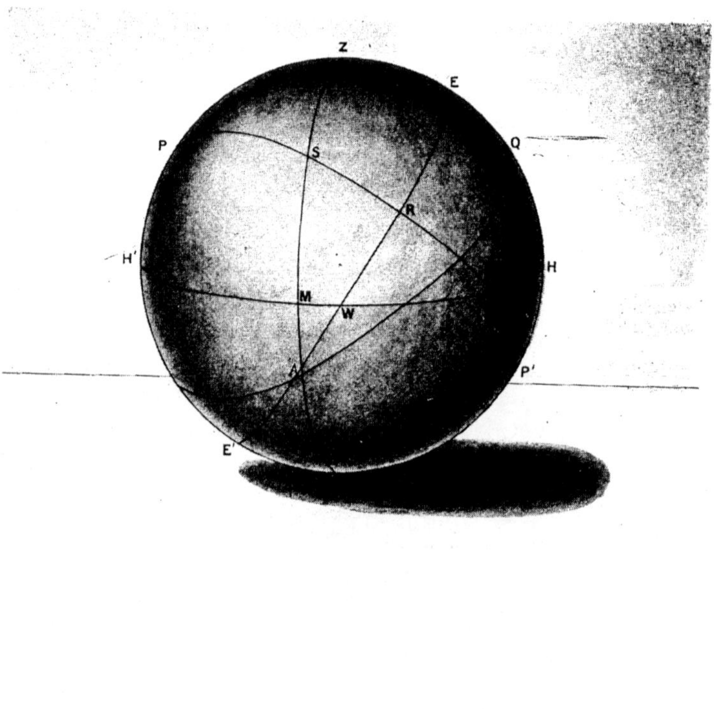

Z
E
P
Q
S
R
H'
H
M
W
A
P'
E'

astronomy in connection with defining the positions of celestial objects. The letters are references to Fig. 1.

The direction of the plumb-line at any place is called the vertical for that place. If the direction of the vertical be produced indefinitely, both up and down, it will intersect the surface of the celestial sphere in two points, called the **zenith** and the **nadir,** respectively (Z and Z').

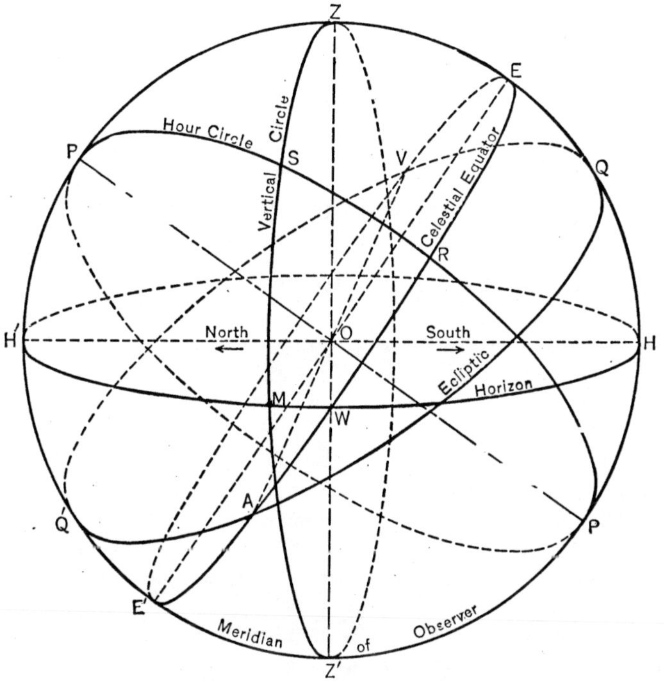

FIG. 1. THE CELESTIAL SPHERE.

A plane through the earth perpendicular to this direction will cut the surface of the celestial sphere in a great circle, called the horizon for that place (HWH').

If the axis of rotation of the earth be produced indefinitely it will pierce the surface of the celestial sphere in the north and the south celestial poles (P and P'). A plane perpendicular to this axis at the center of the earth will cut the surface of the earth in the terrestial equator and the surface of the celestial sphere in the celestial equator ($EVE'A$).

Vertical circles ($ZSMZ'$) are great circles on the surface of the celestial sphere, passing through the zenith and the nadir.

Hour circles ($PSRP'$) are great circles through the celestial poles.

The **meridian** of the observer ($PZP'Z'$) is a great circle through the zenith and the celestial poles. It is at the same time a vertical circle and an hour circle. The projection of this meridian upon the earth is the meridian used in plane surveying.

The **prime vertical** is the vertical circle whose plane is perpendicular to the plane of the meridian. It cuts the horizon in the east and west points. The meridian cuts the horizon at the north and the south.

The great circle on the celestial sphere which the sun's center appears to describe in its (apparent) yearly motion around the earth is called the ecliptic ($Q'VQA$), and the angle which its plane makes with the plane of the equator (about 23° 27′) is called the **obliquity of the** ecliptic. The points at which the ecliptic intersects the equator are called the **equinoxes**. The one at which the sun appears to cross the equator, going northward, about March 21 of each year, is called the **vernal equinox** (V); and the one at which it crosses, going southward, about September 22, is called the **autumnal equinox** (A).

The **latitude** of a place may be defined as the angular distance of the place north or south of the equator; or more exactly, as the angle which the vertical at the place makes with the plane of the equator. Its limiting values are zero and plus or minus ninety degrees. North latitudes are considered plus and south latitudes minus.

The **longitude** of a place is the angular distance (expressed in either degrees or hours) of the place east or west of some arbitrary reference meridian, usually the meridian of Greenwich, England. More exactly, it is the angle between the planes of the reference meridian and the meridian of the place. Its limits are zero and 180° (or 12 hours) east, and zero and 180° (or 12 hours) west.

Fig. 1 illustrates the relative positions of the lines and points defined above.

CHAPTER II

SYSTEMS OF CO-ORDINATES AND THE ASTRONOMICAL TRIANGLE

5. Spherical Co-ordinates. The work of field astronomy requires that we shall be able to define the positions, or more exactly, the directions of the heavenly bodies. For this purpose four systems of spherical co-ordinates are used. These systems have several characteristics in common. They are all systems

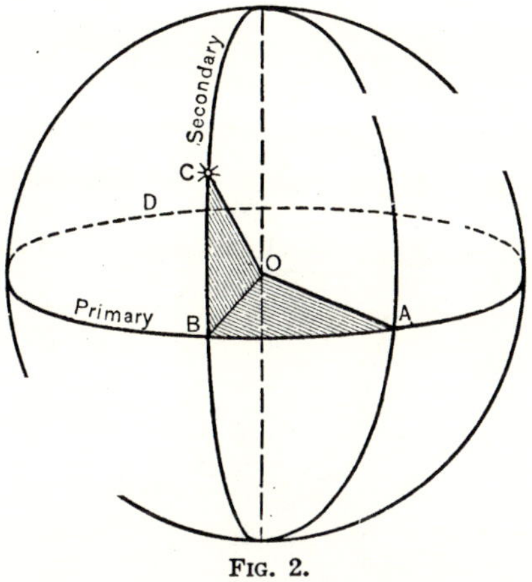

FIG. 2.

of polar co-ordinates, with the earth at the center or pole. In each system the direction of the point in question is located by means of two angles, or arcs. One is measured along a **primary circle** from some starting-point to the foot of a **secondary circle** through the point to be located. The other is measured along the secondary circle from the primary circle to the point to be located. The plane of the secondary circle is always perpendicular to the plane of the primary circle.

In Fig. 2 the direction of point C is determined by the angle AOB, or the arc AB measured from A along the primary circle ABD to B, and the angle BOC, or the arc BC measured along the secondary circle from the primary circle to point *C*. The plane of the arc *BC* is perpendicular to the plane of the arc *AB*. Note that only the direction of the point is determined; no account is taken of its distance—of the length of the radius. This method of locating points is common to the four systems. Of the four we shall use three.

6. The Horizon System. System I, sometimes called the Horizon System, has for its primary circle the horizon, and for

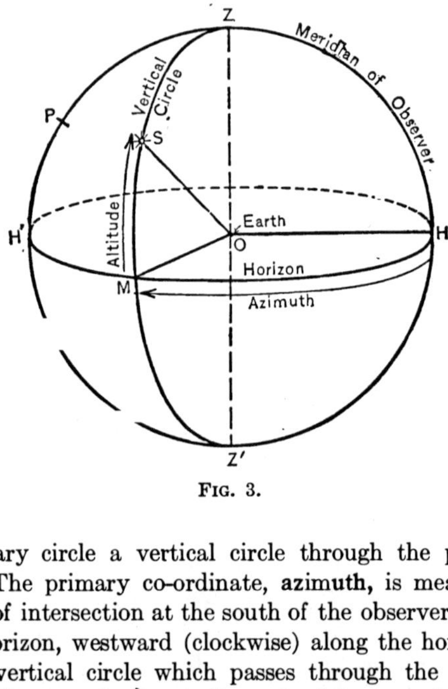

FIG. 3.

its secondary circle a vertical circle through the point to be located. The primary co-ordinate, **azimuth,** is measured from the point of intersection at the south of the observer's meridian, and the horizon, westward (clockwise) along the horizon to the foot of a vertical circle which passes through the point to be located. The secondary co-ordinate is **altitude,** and is measured along the vertical circle from the horizon to the point.

Fig. 3 shows the location of a star, *S*, by its azimuth, *HM*, and its altitude, *MS*.

In some special cases it is more convenient to measure the azimuth from the north instead of from the south, and it is occasionally so measured.

7. The Equator Systems. Systems II and III both have for their primary circle the celestial equator, and for their secondary circle an hour circle through the point to be located.

In System II the primary co-ordinate, **hour angle,** is measured from the point of intersection at the south of the observer's meridian and the celestial equator, westward (clockwise) to the foot of the hour circle through the point. The secondary co-

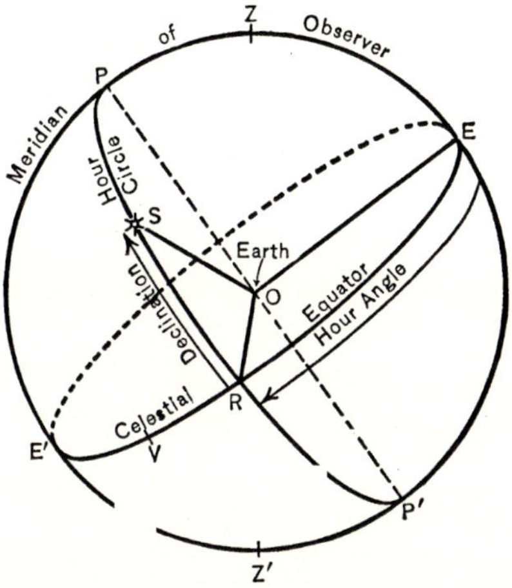

FIG. 4.

ordinate is **declination,** measured along the hour circle from the celestial equator to the point. It is considered plus if measured toward the north celestial pole and minus if toward the south.

In System III the primary co-ordinate is called **right ascension.** It is measured from the vernal equinox eastward (counter-clockwise) along the celestial equator to the foot of the hour circle through the point to be located. The secondary co-ordinate is the same as that of System II. The limiting values of hour angle and of right ascension are in each case 0 hours and 24 hours. The limiting values of declination are zero and plus or minus ninety degrees.

Fig. 4 illustrates the location of a point by System II, and Fig. 5 shows the use of System III.

Some attention should be given to fix in mind the full meaning of the term "hour angle." It is the value in hours (15° per hour) of the spherical angle *EPS* (Fig. 4) or of the arc *ER*. Remembering that the point *S* is traveling westward (clockwise) along a circle whose plane is parallel to the plane of the equator, it will be seen that the hour angle represents the number of hours since the point crossed the meridian of the observer.

For the fourth system of spherical co-ordinates we shall have no use in practical field astronomy.

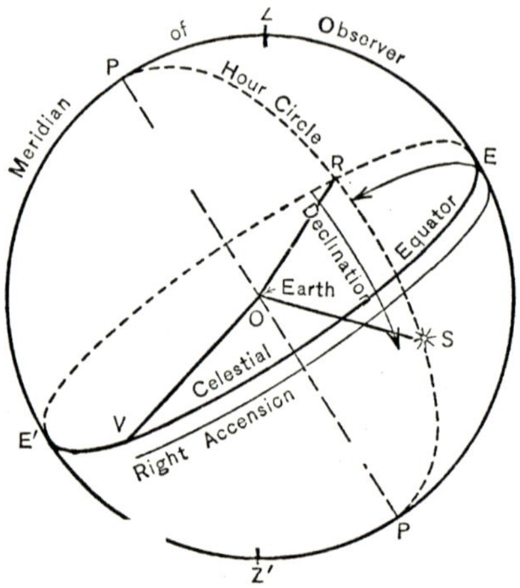

FIG. 5.

Systems I, II, and III are summarized in the following table, and this summary should be thoroughly memorized:

System	Primary Circle	Secondary Circle	PRIMARY CO-ORDINATE				SECONDARY CO-ORDINATE	
			Name	Meas. from	Direction	Limits	Name	Limits
I	Horizon	Vertical Circle	Azimuth	South	Westward	0° to 360°	Altitude	0° to 90°
II	Celestial Equator	Hour Circle	Hour Angle	Intersection at the South of Meridian & Celes. Eq.	Westward	0^h to 24^h	Declination	± 90°
III	Celestial Equator	Hour Circle	Right Ascension	Vernal Equinox.	Eastward	0^h to 24^h	Declination	± 90°

8. Uses of the Three Systems. Each of the three systems possesses certain characteristics in which the others are lacking which give it a place in the work of field astronomy.

System I is the system most used in field measurements of the positions of the heavenly bodies. The reason is that its co-ordinates, the azimuth and altitude of the point in question, may both be measured directly with the engineer's transit. On the other hand, because of the rotation of the earth, the azimuth and altitude of a given point are continually changing; and these co-ordinates also depend on the position of the observer. This system is therefore undesirable for permanent records of the positions of the heavenly bodies.

In System II the first of these difficulties is done away with. The celestial equator, from which the secondary co-ordinate (declination) is measured, lies in a plane normal to the axis of rotation of the earth; and is therefore independent of that rotation. Since the equator is also independent of the observer's position, the declination of a fixed point, such as a fixed star, is independent of the time (*i.e.*, of the rotation of the earth) and of the observer's position. It is very nearly a constant quantity. Any variation may be computed, so that the declination of a heavenly body at a given time may be regarded as a permanent record. The primary co-ordinate of System II (hour angle) increases uniformly with the time, and may therefore be measured with a watch or chronometer.

In System III the point of reference from which the primary co-ordinate is measured shares in the apparent rotation of the celestial sphere, so that the right ascension of a fixed point does not change with time. There are some slight changes in the right ascensions of the fixed stars, due to a slight movement of the vernal equinox. The amounts of these slight variations may be computed; so that the right ascension of a fixed point, once determined, may always be regarded as a known quantity. Since both co-ordinates of this system are independent of the time and of the observer's position and are nearly constant (the amount of any variation being obtainable), they are suitable for use as permanent records of the positions of the heavenly bodies; and they are so used in the "American Ephemeris and Nautical Almanac" and similar works. Here are tabulated the right ascensions and the declinations of the sun, planets, moon, and several hundred of the fixed stars.

9. Relation between the Systems. Since all three systems of co-ordinates have their uses in the work of field astronomy, it is essential that we be able to translate from one system to another. It is in this connection that the assumption that the heavenly bodies are situated on the surface of a celestial sphere comes into play; for if arcs of great circle are considered drawn through the proper points, forming a spherical triangle, most of the problems in transformation of co-ordinates become simply problems in spherical trigonometry. This spherical triangle is always formed (see Fig. 6) by an arc of the meridian of the observer (PZ), an arc of a vertical circle through the point to be located (ZS), and an arc of an hour circle through that point (PS).

This triangle is so much used in the work of field astronomy

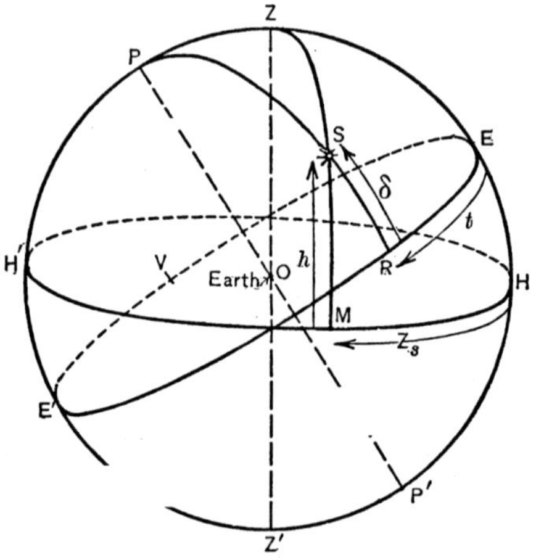

FIG. 6.

that it is called the astronomical triangle; or sometimes, from the letters at its vertices, the "SPZ" triangle.

It is essential that we become thoroughly familiar with the quantities that go to make up the parts of this triangle. In the northern hemisphere the vertices are always at the north celestial pole, the observer's zenith, and the star or other point whose co-ordinates are under consideration.

The angle P at the pole is always the hour angle of the star S if the star is on the western side of the meridian, as shown in Fig. 6; or it is equal to 24 hours minus the hour angle if the star is on the eastern side of the meridian, as in Fig. 7.

The angle Z at the zenith is equal to $(180° - Z_s)$ if the star is on the western side of the meridian, as in Fig. 6; or to $(Z_s - 180°)$ if it is on the eastern side as in Fig. 7.

The angle S at the star is called the parallactic angle. It is very little used.

The arc ZE is, by definition, the observer's latitude; and therefore the arc PZ, or s, is equal to $(90° - \phi)$, and it is called the co-latitude.

The arc MS is the altitude of the star, so that the arc SZ, or p, is equal to $(90° - h)$, and is called the co-altitude, or sometimes the zenith distance.

The arc RS is the declination of the star, and therefore the

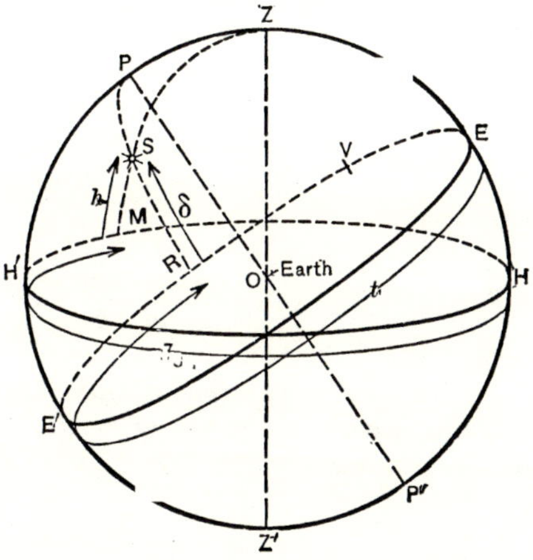

FIG. 7.

arc PS, or z, is equal to $(90° - \delta)$ and is called the co-declination, or sometimes the polar distance.

Thus each part of the astronomical triangle, with the exception of the angle at the star, may be expressed in terms of the observer's position on the earth's surface (latitude) or the co-ordinates of the star.

It may be convenient to note for use in future solutions of the astronomical triangle for hour angle or for azimuth, that if t is less than 12 h, or 180°, Z_s is less than 180°; and if t is greater than 12 h, or 180°, Z_s is greater than 180°.

The values of the five parts of the astronomical triangle defined above are summarized in the following equations:

$$\mathbf{s = 90° - \phi} \quad (14)$$

$$\mathbf{p = 90° - h} \quad (15)$$

$$\mathbf{z = 90° - \delta} \quad (16)$$

$$\mathbf{Z = 180° - Z_s} \quad (17)$$

or

$$\mathbf{= Z_s - 180°} \quad (17a)$$

$$P = t \quad . \quad . \quad . \quad . \quad . \quad . \quad . \quad . \quad . \qquad (18)$$

or

$$= 360° - t \quad . \quad . \quad . \quad . \quad . \quad . \quad . \qquad (18a)$$

10. Relation between Systems I and II. If the latitude of the place is known, and it is required to change from System I to System II, the problem becomes:

Given, in the astronomical triangle: s, p, Z.

Required: P, z.

Three parts of the spherical triangle being given, it may be solved for the two parts required; using equation (1) (from Appendix A), to find the side z, and then Equation (3) to find the angle P.

If it is required to change from System II to System I, the problem is:

Given, in the astronomical triangle: s, z, P.

Required: p, Z.

Again three parts of a spherical triangle are known, and the triangle may be solved for the two required, obtaining first the side p and then the angle Z by the two equations mentioned above.

Solution for Z_s and h from ϕ, t, and δ may be made directly by means of the following formulas from Chauvenet's " Spherical and Practical Astronomy," Vol. I, Article 14:

$$\tan Z_s = \frac{\cos M \,.\, \tan t}{\sin (\phi - M)} \quad . \quad . \quad . \quad . \quad . \qquad (19)$$

$$\tan h = \frac{\cos Z_s}{\tan (\phi - M)} \quad . \quad . \quad . \quad . \quad . \qquad (20)$$

where M is such an angle that:

$$\tan M = \frac{\tan \delta}{\cos t} \quad . \quad . \quad . \quad . \quad . \quad . \quad .$$

11. Relation between Systems II and III. Since in the second and third systems the secondary co-ordinates are the same (*i.e.*, declination in each case), the problem of changing from one system to the other becomes merely one of changing hour angle to right ascension, or vice versa.

Recalling the definitions of these quantities and referring to Fig. 8, we see that the arc ER is the hour angle of the body S, and that the arc VR (V being the vernal equinox) is its right ascension.

The arc EV may be regarded as the hour angle of the vernal equinox, or read in the direction VE, as the right ascension of the meridian. It is evident that the arc ER plus the arc VR is equal to the arc EV. This relation always holds true, no

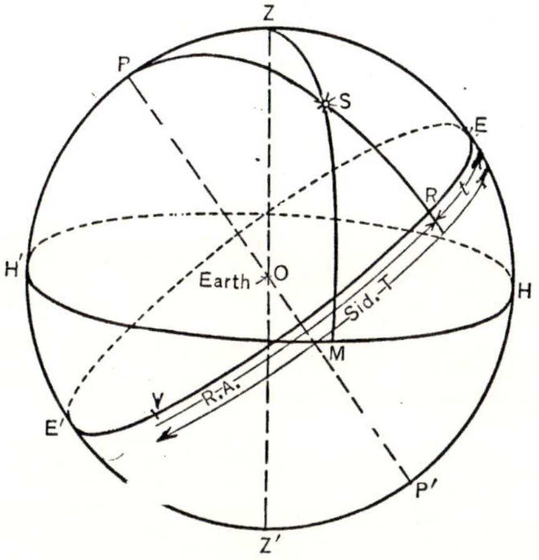

FIG. 8.

matter what the position of the point S may be; provided that when necessary we add 24 hours (or 360°) to the hour angle of the vernal equinox.

It is a very important principle that:

The right ascension plus the hour angle of any body is equal to the hour angle of the vernal equinox, or to the right ascension of the meridian.

We shall learn a little later that the hour angle of the vernal equinox is called "Sidereal Time"; and having studied the measurement of time it will be apparent how this important principle comes into play, not only in the transformation of co-ordinates but in a large share of the problems that we shall have to solve.

If changes between Systems I and III are desired they may be made through the medium of System II.

12. Some Common Solutions of the Astronomical Triangle. Two problems which occur so frequently in the work as to deserve special mention, and which call for solutions of the astronomical triangle, are:

(1) Knowing the latitude, and having given the declination and altitude of a body, to find its hour angle and azimuth.

For computing the angle P of the astronomical triangle, from which the hour angle may be obtained, any one of the following formulas (from Appendix A) may be used:

$$\sin\frac{P}{2} = \sqrt{\frac{\cos(k+\phi)\,.\,\cos(k+\delta)}{\cos\phi\,.\,\cos\delta}} \qquad (5)$$

$$\cos\frac{P}{2} = \sqrt{\frac{\sin k\,.\,\cos(k+h)}{\cos\phi\,.\,\cos\delta}} \quad .\ .\ .\ .\ . \qquad (7)$$

$$\tan\frac{P}{2} = \sqrt{\frac{\cos(k+\phi)\,.\,\cos(k+\delta)}{\sin k\,.\,\cos(k+h)}} \quad .\ . \qquad (9)$$

For computing the angle Z, from which to obtain the azimuth, Z_s or Z_n, any of the following (from Appendix A) may be used:

$$\sin\frac{Z}{2} = \sqrt{\frac{\cos(k+\phi)\,.\,\cos(k+h)}{\cos\phi\,.\,\cos h}} \quad .\ . \qquad (4)$$

$$\cos\frac{Z}{2} = \sqrt{\frac{\sin k\,.\,\cos(k+\delta)}{\cos\phi\,.\,\cos h}} \quad .\ .\ .\ .\ . \qquad (6)$$

$$\tan\frac{Z}{2} = \sqrt{\frac{\cos(k+\phi)\,.\,\cos(k+h)}{\sin k\,.\,\cos(k+\delta)}} \quad .\ . \qquad (8)$$

When selecting a formula from which to determine an angle it should be remembered that if the angle is near 90° the value may be found more accurately through its cosine, while for a small angle the sine gives the greater precision. More precise than either, because of the rapid variation of the function, are the tangent formulas.

(2) The second solution of the astronomical triangle to be noted here concerns circumpolar stars.

If the co-declination or polar distance of a star is less than the latitude of the observer the star will not at any point in its daily rotation go below his horizon; but would, if the light of the sun were not so bright as to obscure it, be always visible. Such a star is called a circumpolar star.

The circle which any circumpolar star appears to follow in its daily motion is at two points tangent to vertical circles. See

Fig. 9. These points are the ones at which the star appears farthest east (at S in Fig. 9) and farthest west (at S'). These two positions are called the points of greatest elongation.

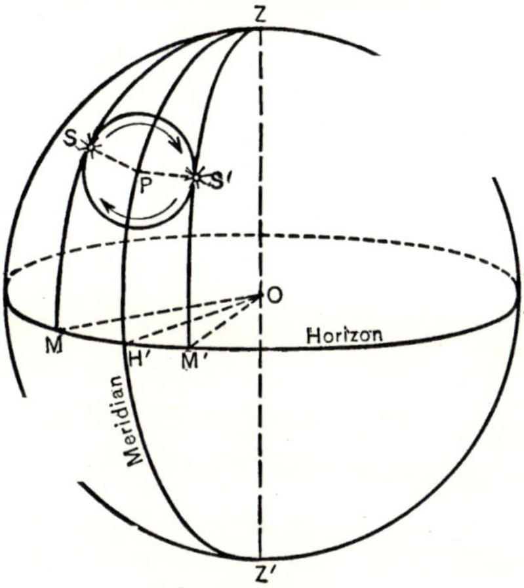

FIG. 9.

In these cases the astronomical triangle is right-angled at S (or S'), and the formulas from which the azimuth and hour angle for this position of the star may be obtained are (from Appendix A):

$$\sin Z = \frac{\cos \delta}{\cos \phi} \quad \ldots \ldots \quad (12)$$

$$\cos P = \frac{\tan \phi}{\tan \delta} \quad \ldots \ldots \quad (13)$$

CHAPTER III

MEASUREMENT OF TIME

13. The Unit of Measurement. The unit of measurement of time is based upon the period of rotation of the earth about its axis. It is not known that this period is invariable; but the variations, if any exist, are too small to be measured, and the rotation is considered uniform. The differences between the several methods of measuring time arise from the different methods of counting these rotations. This much is common to all: that the counting of the periods of rotation is done by noting successive passages of some reference point over the meridian of an observer.

Every point on the celestial sphere crosses any given meridian twice during each period of rotation of the earth. The instant when any point is on the same half of the meridian as the zenith is called the **upper transit** or **upper culmination** of the point. The instant when it is on the opposite half of the meridian is called the **lower transit** or **lower culmination** of the point. Except in the case of the circumpolar stars, which never go below the horizon, the upper transit is the only one visible; and unless otherwise stated it is the one that is meant when the transit of a body is mentioned.

14. Apparent Solar Time. The most common methods of measuring time are based on the use of the sun as a reference point in counting the rotations of the earth. The interval between two successive upper transits of the sun's center over the meridian of an observer is called an **apparent solar day,** and the system of measurement of which this interval is the unit is called **apparent solar time.** The instant of transit at any place is **apparent noon** for that place, and the apparent solar time at any instant is the hour angle of the sun's center at that instant; *i.e.*, the number of hours since the sun's center crossed the meridian. It is the time as given by a sun-dial.

Because of the earth's motion around the sun in the same direction as its rotation about its own axis, the direction of the reference point is continually changing; and the interval between two successive transits of the sun's center is not the true period of rotation of the earth. Moreover, since the rate of motion of the earth in its orbit is not constant, the change in direction of the reference point is not constant; and therefore the lengths of apparent solar days are different at different times of the year.

15. Mean Solar Time. To avoid the inconvenience of a unit of variable length, use is made of the convention of a fictitious **mean sun.** This fictitious "sun" is assumed to have a motion around the earth, the summation of which amounts in a year to exactly the same as the apparent motion of the real sun—to one complete revolution. The essential difference is that while the apparent motion of the real sun takes place along the ecliptic at a varying rate, the assumed motion of the mean sun takes place along the celestial equator at a constant rate.

A **mean solar day** is the interval between two successive transits of the mean sun over the same meridian. It is a constant unit and is equal in length to the average of all the apparent solar days in the year.

Mean noon at any place is the instant of upper transit of the mean sun at that place.

The **mean solar time** at any place is the hour angle of the mean sun at that place at the given instant.

16. Relation between Apparent and Mean Solar Time—The Equation of Time. The two chief causes of the irregularity of the apparent motion of the sun, which in turn causes the difference between apparent and mean solar time, are as suggested above: the variable rate of motion of the earth in its orbit around the sun, and the inclination of the plane of this orbit to the plane of the equator.

The earth's orbit is elliptical in shape, and in order that the earth may obey the laws of gravitation in its motion—*viz.*, that the line joining it at any point in its path to the sun shall sweep over equal areas during equal intervals of time—it is necessary that its rates of motion at different times should be different. In the winter, when the sun is nearest the earth, the rate of angular motion of the "radius" is faster; and therefore the apparent solar days are longer at that time of the year. In the summer, for a similar reason, they are shorter.* The maxi-

* The relative length of apparent solar days at different times of the year is not to be confused with the relative length of the periods of daylight at these seasons. The inclination of the earth's axis of rotation to the plane of its orbit is such that in the winter when the earth is actually nearest the sun the latter has its greatest southern declination; *i.e.*, it is farthest south of the equator, so that the period between sunrise and sunset in the northern hemisphere is shortest. Since the rays from the sun strike the surface of the northern hemisphere so obliquely at that time, we have our coldest weather.

mum difference between apparent and mean time due to this cause alone is about eight minutes, plus or minus. Due to the second cause mentioned there may be a maximum difference of about ten minutes.

The maximum combined effect of these two causes is a little over sixteen minutes; and this difference between apparent and mean time, varying in amount from zero to the maximum, is called the **equation of time.** It is continually changing, but its value for any given instant may be computed. Data for this computation are found in the "American Ephemeris and Nautical Almanac," to which reference has already been made, and which will be described more fully in Chapter IV.

To change from apparent to mean time or vice versa, it is

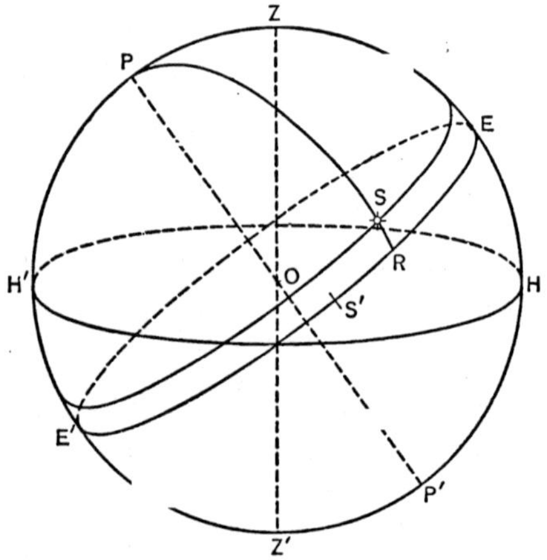

FIG. 10.

only necessary to add or to subtract the equation of time for the instant at which the change is to be made, addition or subtraction being determined by the time of year and the direction of change.

Reference to Fig. 10 may help to make clear the process of changing from apparent to mean solar time, or vice versa.

Let S and S' represent the real sun and the fictitious or mean sun, respectively. Both are traveling westward, clockwise, in their daily rotation—the mean sun on the equator and the real sun on a path which is above and nearly parallel to the equator, and which might be likened to one turn of a spiral or to one turn of a helical spring.

Local apparent solar time—the hour angle of the real sun, the number of hours since the real sun crossed the meridian—is the arc ER. Local mean solar time—the hour angle of the mean sun, the number of hours since the mean sun crossed the meridian—is the arc ES'. Either the mean sun or the real sun might be in advance, depending on the time of year. The equation of time, the amount by which one is in advance of the other, is the arc RS'.

Since we are able to compute the value of the equation of time for any desired instant, it is obvious how if either local apparent time or local mean time is known we may obtain the other.

The determination of the instant of time in one system of measurement corresponding to a given instant in another system is always a matter of comparing the positions of the reference points in the two systems. Stated in other words, the problem is: Having given the hour angle of the reference point of one system, it is required to find the hour angle of the reference point of the other system.

Examples of changing from apparent to mean time and vice versa will be given after we have learned more about the "American Ephemeris and Nautical Almanac,"—in Chapter V.

17. Astronomical and Civil Time. For astronomical purposes the mean solar day is divided into twenty-four hours, beginning at the instant of mean noon; and the hours are subdivided into minutes and seconds.

For ordinary purposes the mean solar day is divided into two periods of twelve hours each: P.M. (*post meridiem*), beginning at mean noon and continuing until twelve o'clock, midnight; and A.M. (*ante meridiem*), from midnight until mean noon of the next day. The **civil day** is considered to extend from midnight to midnight. The **astronomical day** begins at mean noon on the civil day of the same date.

Astronomical time as well as civil time may be either apparent or mean.

For changing from one scheme of division of the solar day to the other the following rules may be used:

To change Astronomical Time to Civil Time:

If less than 12 hours call it P.M.

If greater than 12 hours subtract twelve hours, add one day to the date, and call it A.M.

To change Civil Time to Astronomical Time:

If A.M. add 12 hours, drop one day from the date and drop the A.M.

If P.M. drop the P.M.

For instance:

July 6, 8 h., astronomical time = July 6, 8 P.M., civil time.

May 11, 4 A.M., civil time = May 10, 16 h., astronomical time.

18. Standard Time. Since local mean solar time at any instant is the hour angle of the mean sun at that instant, it is evident that all places not on the same meridian will have different local mean times. To avoid confusion from this source, a uniform system of time was established in the United States in 1883. The country was divided into belts, each 15° or one hour of longitude wide, each belt to use as **standard time** the local mean time of a central meridian. There are four such belts across the country, using the times of the 75th, 90th, 105th, and 120th meridians (west of Greenwich). Local mean time of the 75th meridian is called **Eastern Time**; of the 90th, **Central Time**; of the 105th, **Mountain Time**; of the 120th, **Pacific Time.** Some of the eastern provinces of Canada use the local mean time of the 60th meridian, called **Atlantic Time.**

The theoretical boundaries of these time belts have been shifted to suit local convenience, and the boundaries now depend largely on the location of cities and railroads; but the difference in time between one belt and the next is always one hour.

For instance, when it is noon at Greenwich, England, it is 7 A.M. by Eastern Time, 6 A.M. by Central Time, 5 A.M. by Mountain Time, and 4 A.M. by Pacific Time.

To change from Standard Time to Local Mean Time at any place:

Express the difference in longitude between the standard and local meridians in units of time, and if the place is east of the standard meridian add this difference to the standard time; if west, subtract.

To change from Local Mean Time to Standard Time at any place:

The procedure is exactly the reverse.

To determine whether to add or to subtract a "correction," it is only necessary to remember that: **The farther east a place is the later it is by local mean time at any instant. Standard time is simply local mean time at a standard meridian.**

Examples of the conversion of local mean to standard time, and vice versa, are given in Chapter V.

19. Sidereal Time. So far we have studied methods of measuring time which are based on the use of the sun as a reference point for counting the rotations of the earth about its axis —*i.e.*, solar time—and these are the methods in common use for most purposes. But since the sun appears to make one revolution a year around the earth, causing successive transits of the sun's center over the same meridian to occur at slightly different points in the earth's rotation, the intervals between successive transits are not a true measure of the period of that rotation.

For astronomical purposes a more precise determination is needed, and sidereal time is used. The **sidereal day** is the interval between two successive upper transits of the vernal equinox over the same meridian, and the **sidereal time** at any place is the hour angle of the vernal equinox at that place at the given instant.

If the vernal equinox were a fixed point the sidereal day would be an exact measure of the period of the earth's rotation. The equinox has a slow westward movement, but it is so slight that the length of a sidereal day differs from the true period of one rotation by only about $0^s.01$; and as sidereal time is not used over long intervals (dates are always kept in solar time) cumulative errors are avoided; and the sidereal day as defined above is the one actually used, without correction.

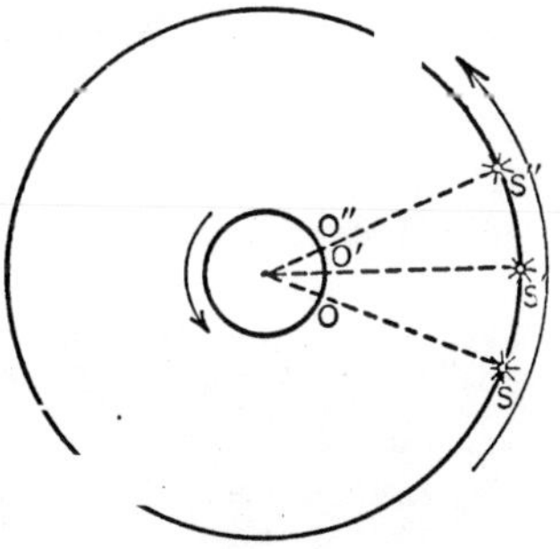

FIG. 11.

20. Relation between Sidereal and Mean Solar Intervals of Time. It has been mentioned that to the apparent motion of the sun around the earth is due the difference between sidereal and solar time. Reference to Fig. 11 may help to make this relation between the two more clear. Let the large circle represent the celestial equator, along which the mean sun is assumed to move at a uniform rate; and let the small circle represent the earth. Let an observer be at O on the earth when the sun is on his meridian at S. Now, while the earth is making one revolution in the direction indicated by the arrow, the mean sun is also moving in the same direction; so that its next transit over the observer's meridian takes place when the observer's position has revolved through nearly 361° to O', and the mean sun is at S'. At the next transit the observer is at O'' and the mean sun at S''. In one year the sun has apparently completed one revolution around the earth and is again at S, so that in that time one rotation of the earth has not been counted. In the meantime the vernal equinox, having remained practically a fixed point, has registered the exact number of rotations that have taken place. There are, therefore, one less solar than sidereal days in a year.

The length of the "tropical year"* has been determined to be 365.2422 mean solar days, and since there are one more sidereal than solar days the relation between the two is given by the following equations:

$$\mathbf{365.2422}\text{ mean solar days} = \mathbf{366.2422}\text{ sidereal days} \quad (21)$$

$$\mathbf{1}\text{ mean solar day} = \mathbf{1.0027379}\text{ sidereal days} \quad (22)$$

$$\mathbf{1}\text{ sidereal day} = \mathbf{0.9972696}\text{ mean solar days} \quad (23)$$

Tables I and II are arranged for conversion of sidereal into mean solar intervals, and vice versa; and are more convenient to use than formulas.

Tables II and III of the "American Ephemeris and Nautical Almanac" (near the end of the book) are for the same purpose.

It will be convenient to remember that a mean solar hour is about ten seconds longer than a sidereal hour, and that a mean solar day is about 3 min. 56 sec. longer than a sidereal day.

It must be kept clearly in mind that the difference between the units of the two systems—sidereal and solar—is simply this: A sidereal hour as a unit of time is the interval of time required for the vernal equinox to pass over 15° or one "hour" of arc. A mean solar hour as a unit of time is the interval of time required for the mean sun to pass over 15° or one "hour" of arc. The difference arises solely from the different rates of apparent motion of the two reference points. It is a difference between two intervals of time—not between two unit angles or arcs. That is, an hour angle, for instance, is not measured in "sidereal hours" or "solar hours" but simply in "hours"—twenty-fourths of a circumference—or in degrees, 15° per hour.

* The tropical year is the interval of time between two successive passages of the sun through the vernal equinox.

It must also be carefully noted that the relation between solar and sidereal intervals of time, as discussed in this article, is simply a comparison of units. It must not be confused with the relation between mean solar and sidereal time at a given instant, discussed in the next article. The one has to do with the relative sizes of the units of measurement, the other concerns the relative positions of the mean sun and the vernal equinox at a given instant.

21. Relation between Sidereal and Mean Solar Time at a Given Instant. We have learned (Art. 11, page 12) that the right ascension of any body plus its hour angle is equal to the hour angle of the vernal equinox, and (Art. 19, page 21) that the hour angle of the vernal equinox is sidereal time. It follows, then, that:

The sidereal time at any instant is equal to the right ascension plus the hour angle of any body at that instant, or

$$\text{Sid. T} = \text{R A} + \text{t} \quad . \quad . \quad . \quad . \quad . \quad (24)$$

This is probably the most used and perhaps the most important of all the equations needed in field astronomy.

If the mean sun is the body under consideration the following equation is true:

Sidereal Time = Right Ascension of Mean Sun + Hour Angle of Mean Sun.

But the hour angle of the mean sun is local mean solar time, therefore:

$$\text{Sid. T} = \text{A} + \text{LMT} \quad . \quad . \quad . \quad . \quad . \quad (25)$$

$$\text{LMT} = \text{Sid. T} - \text{A} \quad . \quad . \quad . \quad . \quad (26)$$

Granting that we can obtain the value of A for the desired instant, these equations will enable us to obtain the sidereal time corresponding to any given instant of local mean solar time, or to find the local mean solar time corresponding to any given instant of sidereal time.

The right ascension of the mean sun is entirely independent of the location of the observer, and is dependent only on the absolute instant of time. At some instant about March 22 of each year the mean sun is at the vernal equinox, and its right ascension is zero. Leaving the vernal equinox, it moves eastward along the celestial equator at a constant rate; and therefore A is equal to zero at some instant about March 22 of each year and increases, constantly, to 24 hours (or 0 hours) at some instant on March 22 of the next year. The value of A for the instant of mean noon at Greenwich for each day in the year is given in the "American Ephemeris and Nautical Almanac." To find the value of A for any instant, it is necessary to find the interval of time that has elapsed since the last preceding mean noon at Greenwich, determine the increase in the right ascension of the mean sun during that interval, and add that increase to the value taken from the tables for the instant of Greenwich mean noon preceding.

This constant increase in the right ascension of the mean sun is simply the gain of the apparent motion of the vernal equinox over that of the sun during the interval considered, and it is therefore equal to the difference between sidereal and mean solar time for that interval. In other words, it is the difference between the number of sidereal units in the interval of time and the number of mean solar units in the same interval. This increase, or "correction," may therefore be taken directly from Tables I and II at the back of this book or from Tables II and III at the back of the "Nautical Almanac."

Problems in changing from sidereal to mean solar time and from mean solar to sidereal time will be solved in Chapter V.

Changes from apparent solar time to sidereal time and vice versa may be made by first changing to mean solar time in each case.

CHAPTER IV

THE AMERICAN EPHEMERIS AND NAUTICAL ALMANAC

22. The Ephemeris. The "American Ephemeris and Nautical Almanac" is published yearly, three years in advance, by the Nautical Almanac Office of the United States Naval Observatory, at Washington, D. C., and is sold by the Superintendent of Documents. (See note on page 28.) It contains data of use to surveyors, navigators, and others for astronomical calculations, made up from the results of observations with large instruments at the principal observatories, and from calculations.

These data comprise ephemerides* of the sun, moon, planets, and stars, the equation of time, semi-diameters, and horizontal parallaxes of heavenly bodies, convenient tables for conversion of units, etc., as well as a great deal of data in regard to eclipses and other phenomena of more interest to astronomers than to surveyors. Many of these quantities vary with the time, so their values are given for regular intervals of time for the meridian of Greenwich or of Washington.

The Almanac is divided into three parts. Part I is an "Ephemeris for the Meridian of Greenwich." The data in Part I of most interest to surveyors is the ephemeris of the sun. There is also an ephemeris of the moon which may be of occasional use. In the ephemeris of the sun there are given *for the instant of mean noon at Greenwich* for each day in the year: The sun's apparent right ascension and declination with the hourly variation in each, its semi-diameter and horizontal parallax, the equation of time with its hourly variation and with the proper algebraic sign for changing from mean to apparent time (the opposite sign would be used in changing from apparent to mean), the sidereal time, or right ascension of the mean sun. Note that this last quantity is the sidereal time at Greenwich mean noon, or the right ascension of the mean sun at Greenwich mean noon; *i. e.*, at the instant when the mean sun is on the meridian of Greenwich. All the above-mentioned data are tabulated on successive left-

*** By "ephemeris" is meant a catalogue of the positions of a celestial body at equidistant intervals of time, usually given in terms of the body's right ascension and declination.**

hand pages, beginning on page 2. On the right-hand pages are given other data for the corresponding dates; but only the last column, which gives the mean time of sidereal noon for each day (*i.e.*, the Greenwich mean time when the vernal equinox is on the meridian of Greenwich), is likely to be of use to the surveyor.

Of the data in Part II, "Ephemeris for the Meridian of Washington," the following are of use to the surveyor: An ephemeris of the sun for the instant of Washington apparent noon for each day, ephemerides of thirty-five circumpolar stars for the instant of upper transit at Washington on each day, and ephemerides of 825 other stars for the instant of upper transit at Washington at intervals of ten days.

The ephemeris of the sun gives the right ascension and declination of the sun at the instant of Washington apparent noon with their hourly variations, the equation of time with its hourly variation and with the proper algebraic sign for changing from apparent to mean time, the semi-diameter of the sun, the sidereal time required for the semi-diameter of the sun to pass the meridian, and the sidereal time of mean noon at Washington for each day in the year.

The table of which each page is headed, "Apparent Places of Stars, 19—, Circumpolar Stars," gives for the time of upper transit at Washington for each day in the year the right ascension and declination of thirty-five circumpolar stars. The table of which each page is headed, "Apparent Places of Stars, 19—," gives for the time of upper transit at Washington at intervals of ten days the right ascension and declination of each of 825 other stars.

It will be seen upon examination of the tables that the right ascensions and declinations of these stars change so slowly that, though they have been computed for the instants of upper transit of the stars at Washington, there will be no appreciable error in any work for which a surveyor is likely to need the data if these values are used for the upper transit on the corresponding date at any place in the United States. The dates in the column at the left of each page of the star ephemerides are in mean solar time; and each contains a decimal, as April 3.7, July 17.1, etc. The decimal part of the date indicates the approximate time, in tenths of twenty-four hours, from mean noon of the day indicated by the integral part of the date to the time of upper transit of the star.

Part III, "Phenomena," is of no direct use in practical field astronomy.

Following Part III is a table giving the latitude and longitude of (in 1916) 252 places on the earth's surface, and tables numbered I to VII, for the convenient conversion of units, etc. Following the tables, and immediately preceding the general index, is a useful "Index to Apparent Places of Stars."

A comparatively small, paper-covered book, called the "American Nautical Almanac," is also published by the Nautical Almanac Office. (See note on page 28.) It contains, arranged in slightly different form from that in which the same data are given in the larger book, tables which are sufficient for the work of practical field astronomy; though it is often convenient in preparing for observations for time to have the longer lists of stars from which to choose. The arrangement and use of the tables will be understood upon examination, and need no explanation here.

Reprints of portions of the "Nautical Almanac" which are published by the different instrument-makers are useful chiefly in connection with the solar attachments for transits. The use of these attachments is discussed briefly in Appendix B.

23. Interpolation. If the right ascension or the declination of the sun or the equation of time is desired for a given instant of local mean time it is necessary to add (algebraically) to the corresponding quantity given in the tables for the instant of Greenwich mean noon preceding, the hourly change or variation multiplied by the number of hours that have elapsed between the instant of Greenwich mean noon and the instant for which the quantity is desired. The interval between Greenwich noon and local noon is equal to the longitude of the place expressed in units of time. The interval between Greenwich noon and noon by standard time is equal to the longitude of the standard meridian, and therefore to some number of whole hours.

If the sun's right ascension or declination or the equation of time at some instant of local apparent time is desired it may be obtained most exactly from the ephemeris for Washington apparent noon in a manner similar to that outlined above. The interval between Washington noon and local noon is equal to the difference, expressed in units of time, between the longitude of Washington ($5^h\ 08^m\ 15^s.78$ west of Greenwich) and the longitude of the observer.

The work of making these interpolations is illustrated in the course of the solution of the problems in conversion of time, given in Chapter V.

It is usually assumed that the rate of variation of any quantity is constant between any two tabular values. This is not always quite true, however; and the variations per hour are not, in general, tabular differences but rates of change for the instants for which they are given—differential coefficients. Somewhat more precise results may therefore be obtained by interpolating from the nearer of two tabular quantities.

NOTE.—The "American Ephemeris and Nautical Almanac" and the "American Nautical Almanac" are two of the "Astronomical Papers" which in turn constitute a part of the Public Documents of the United States. As is true of the greater number of Public Documents, they may be obtained from the Superintendent of Documents, payment in advance being required. The price charged is only enough to cover the cost of printing, binding, paper, etc. The "American Ephemeris and Nautical Almanac" is sold for one dollar, and the "American Nautical Almanac" for thirty cents. Price List 57, giving the titles and prices of all the Astronomical Papers, will be sent free on request by the Superintendent of Documents.

The following instructions are copied from that list:

"Remittances should be made to the Superintendent of Documents, Government Printing Office, Washington, D. C., by postal money-order, express-order, or New York draft. If currency is sent, it will be at sender's risk.

"Postage stamps, coins defaced or worn smooth, foreign money, and uncertified checks will not be accepted.

"No charge is made for postage on documents forwarded to points in the United States, Alaska, Guam, Hawaii, Philippine Islands, Porto Rico, Samoa, or to Canada, Cuba, Mexico, or Shanghai. To other countries the regular rate of postage is charged."

CHAPTER V

PROBLEMS IN CONVERSION OF TIME

24. To Change Local Mean to Local Apparent Time.

Problem: What was the local apparent time at a place whose longitude is 91° 31′ 30″ W when the local mean time was 8 o'clock A.M., on March 4, 1916?

Read Art. 16, page 17.

Solution: We must first find the equation of time, *E*, for the instant of 8:00:00.0 A.M., *LMT* in longitude 91° 31′ 30″ W on March 4. The first step is to change the civil time to astronomical time. Using the rules given in Art. 17, page 19, it is found that the instant of astronomical time corresponding to March 4, 8:00:00.0 A.M. is March 3, 20:00:00.0. The second step is to find the interval that has elapsed since the last preceding mean noon at Greenwich. Twenty hours have elapsed since mean noon of March 3 at a place $6^h\ 06^m\ 06^s$ (91° 31′ 30″ ÷ 15) west of Greenwich; and therefore $26^h\ 06^m\ 06^s$ have elapsed since mean noon of March 3 at Greenwich, or $2^h\ 06^m\ 06^s$ since mean noon of March 4.

The equation of time for the instant of mean noon of March 4 at Greenwich (taken from page 4 of the "American Ephemeris and Nautical Almanac") is $11^m\ 53^s.64$; and the negative sign indicates that the difference between apparent and mean time is negative; that is, that the amount of the equation of time must be subtracted from mean time to obtain apparent time. Comparing with the value for the preceding day, we see that the equation of time is decreasing, numerically; so that the product of the hourly variation, $0^s.546$, by 2.10 hours since noon is to be subtracted from the value at noon.

The equation of time for the instant at which the change is to be made, then, is: $11^m\ 53^s.64 - 2.10 \times 0^s.546 = 11^m\ 52^s.5$, to be subtracted from mean time. The local apparent time is, therefore, $20^h\ 00^m\ 00^s.0 - 11^m\ 52^s.5 = 19^h\ 48^m\ 07^s.5$, on March 3, 1916.

The following is a convenient form for the solution of this problem:

1916		
LMT, 91:31:30 W, civil, March 4........A.M.		$8^h\ 00^m\ 00^s.0$
LMT, 91:31:30 W, astronomical, March 3....		20 00 00 .0
Longitude west of Greenwich......	6 06 06	
Since Greenwich mean noon, Mar. 4.	2 06 06	
$E = -(11^m\ 53^s.64 - 2.10 \times 0^s.546) = \ldots\ldots$		$-$ 11 52 .5
LAT, 91:31:30 W, March 3.................		$19^h\ 48^m\ 07^s.5$

25. To Change Local Apparent to Local Mean Time.

Problem: What was the local mean time at a place whose longitude is 91° 31′ 30″ W when the local apparent time was $17^h\ 46^m\ 09^s.4$ on March 8, 1916?

Read Art. 16, page 17.

Solution: The same process is followed as in the preceding problem except that the equation of time may be taken from the Washington ephemeris, where it is given for the instant of Washington apparent noon of each day.

The longitude west of Greenwich, $6^h\ 06^m\ 06^s$, minus $5^h\ 08^m\ 15^s.78$ (the longitude of Washington west of Greenwich) gives the longitude of the place as $0^h\ 57^m\ 50^s.2$ west of Washington. The equation of time for an instant which is $17^h\ 46^m\ 09^s.4 + 00^h\ 57^m\ 50^s.2 = 18^h\ 43^m\ 59^s.6$ after the instant of Washington apparent noon on March 8 may best be obtained from the value of the equation of time for March 9, given on page 515 of the "American Ephemeris and Nautical Almanac." (We interpolate from the value for March 9 rather than from the one for March 8 because the former is the nearer.) It will be equal to $10^m\ 39^s.35 + 5.27 \times 0^s.636 = 10^m\ 42^s.7$; and the positive sign indicates that the difference between mean and apparent time is positive, that is, that the amount of the equation of time must be added to apparent time to obtain mean time. (This relation is shown by the heading of the column: "Mean − App.")

Had we used the Greenwich ephemeris and assumed the value of the equation of time as given for mean noon of March 9 to be the proper value for apparent noon of the same day, we should have obtained (following the method of the preceding article and interpolating from the nearer of the two values) the value of the equation of time to be: $10^m\ 42^s.8$. In this case the error is $0^s.1$, and in no case is it likely to be greater than about $0^s.2$—an amount that is of little account in work done with ordinary field instruments. This error could be reduced by a second computation to an amount negligible in any work of field astronomy.

In the "American Nautical Almanac" (the small paper-covered edition) no ephemeris for apparent noon for any place is given, but it is seen from the example above that the one for mean noon may be used as for apparent noon without appreciable error.

The following form of computation is convenient for the solution of the problem above:

	1916		
LAT, 91:31:30 W, March 8................			$17^h\ 46^m\ 09^s.4$
Longitude west of Washington...	00 57 50.2		
Since Wash. app. noon, Mar. 8...	18 43 59.6		
$E = + (10^m\ 39^s.35 + 5.27 \times 0^s.636) =$......		+	10 42 .7
LMT, 91:31:30 W, astronomical, March 8....			$17^h\ 56^m\ 52^s.1$
LMT, 91:31:30 W, civil, March 9........A.M.			5 56 52 .1

26. To Change Standard to Local Mean Time.

Problem: What is the local mean time in longitude 88° 30′ 00″ W when the standard time is 7:31:15 A.M.?

Read Art. 18, page 20.

Solution: The meridian at which local mean time is required is 1° 30′ 00″ east of the standard meridian for the Central time belt. Therefore, following the rules of Art. 18, we must add the difference in longitude, expressed in units of time, to standard time to get local mean time.

Std. T (Central).........................A.M.		$7^h\ 31^m\ 15^s$
Longitude correction, standard to local........	+	6 00
LMT, 88:30:00 W.......................A.M.		$7^h\ 37^m\ 15^s$

27. To Change Local Mean to Standard Time.

Problem: What is the standard (Central) time in longitude 91° 31′ 30″ W when the local mean time is 8:15:27 P.M.?

Solution: The meridian at which the local mean time is given is 1° 31′ 30″, or $0^h\ 06^m\ 06^s$, west of the standard meridian for the Central time belt. Therefore, following the rules of Art. 18, we must add the difference in longitude, expressed in units of time, to the local mean time to get standard time.

LMT, 91:31:30 W.........................P.M.		$8^h\ 15^m\ 27^s$
Longitude correction, local to standard........	+	6 06
Standard time (Central)..................P.M.		$8^h\ 21^m\ 33^s$

28. To Change Standard to Local Apparent Time.

Problem: What was the local apparent time at a place whose longitude is 91° 31′ 30″ W when it was 8:00:00.0 A.M. by standard time (Central), on August 27, 1915?

Read Arts. 16 to 18, pages 17 to 20.

Solution: The procedure is almost exactly that used in Art. 24, page 29. The reason for the slight difference will be self-evident if we remember that standard time in longitude 91° 31′ 30″ W is simply the local mean time at longitude 90° 00′ 00″ W.

1915

Std. T, 91:31:30 W, civil, August 27...... A.M.			$8^h\ 00^m\ 00^s.0$
LMT, 90:00:00 W, astronomical, August 26..			20 00 00 .0
LMT, 91:31:30 W, astronomical, August 26..			19 53 54 .0
Longitude west of Greenwich.......	6 06 06		
Since Greenwich mean noon, Aug. 27.	2 00 00		
$E = -(1^m\ 42^s.02 - 2.00 \times 0^s.704) =$		−	1 40 .6
LAT, 91:31:30 W, August 26..............			$19^h\ 52^m\ 13^s.4$

29. To Change Local Apparent to Standard Time.

Problem: What was the standard (Central) time when the local apparent time at a place whose longitude is 91° 31′ 30″ W, was $16^h\ 08^m\ 17^s.4$ on February 2, 1916?

Read Arts. 16 to 18, pages 17 and 20.

Solution: The procedure is just the reverse of that of the preceding article, with the exception that the equation of time is obtained from the ephemeris of the sun for Washington apparent noon. Interpolation is made from the nearer of two tabular values of the equation of time.

1916

LAT, 91:31:30 W, February 2.............			$16^h\ 08^m\ 17^s.4$
Longitude west of Washington....	0 57 50.2		
Since Wash. app. noon, Feb. 2...	17 06 07.6		
$E = +(13^m\ 53^s.48 - 6.90 \times 0^s.297) =$		+	13 51 .4
LMT, 91:31:30 W, astronomical, February 2..			$16^h\ 22^m\ 08^s.8$
LMT, 90:00:00 W, astronomical, February 2..			16 28 14 .8
Std. T, Central, February 3............. A.M.			4 28 14 .8

30. To Change Local Mean Solar to Sidereal Time.

Problem: What was the sidereal time at 6:00:00.0 A.M., local mean time, in longitude 91° 31′ 30″ W on October 20, 1915?

Read Art. 21, page 23.

Solution: Since the increase in the value of the right ascension of the mean sun during any interval is equal to the gain of the apparent movement of the vernal equinox over that of the mean sun during that interval—*i.e.*, it is equal to the difference between the number of sidereal units and the number of solar units in the interval—it is convenient in working problems to apply this increase in the following manner:

Determine the value of the right ascension of the mean sun for the instant of local mean noon. This quantity will be called A_n. It may be found by adding to the value of the sun's right ascension, A, for the instant of Greenwich mean noon of the same date (*i.e.*, in west longitudes) the increase in right ascension during a solar interval equal to the longitude of the place. This increase may be taken directly from Table II at the back of this book or from Table III in the "Nautical Almanac." For any given longitude this increase or correction for reducing the sun's right ascension at Greenwich mean noon to its value for local mean noon is a constant which may be used for all similar problems. It is the amount by which the distance traveled by the vernal equinox (expressed in hours) exceeds the distance traveled by the sun during the interval that the sun has occupied in coming from the Greenwich meridian to the local meridian.

To this value of A_n add the local mean time expressed in sidereal units, *i.e.*, the sidereal interval since mean noon—the distance expressed in hours that the vernal equinox has traveled since mean noon. This change from solar to sidereal units may also be made by aid of Table II.

This sum—the right ascension of the sun at local mean noon plus the sidereal interval since local mean noon—will be the sidereal time.

For the problem above: First change 6:00:00.0 A.M., October 20 to $18^h\,00^m\,00^s.0$, October 19. The value of A for Greenwich mean noon of October 19 is (from the "American Ephemeris and Nautical Almanac") $13^h\,47^m\,30^s.95$. The increase for $6^h\,06^m\,06^s$ west longitude is, from Table II, $1^m\,00^s.15$ (or, from Table III of the "Nautical Almanac," $1^m\,00^s.140$), making A_n equal to $13^h\,48^m\,31^s.1$. $18^h\,00^m\,00^s.0$ *LMT* expressed in sidereal units is (from Table II, or Table III of the "Nautical Almanac") $18^h\ 02^m\ 57^s.4$. This is the sidereal interval since local mean noon, or the number of hours of arc that the vernal equinox covered while the mean sun was passing over $18^h\ 00^m\ 00^s.0$ of arc.

The sidereal time is therefore $13^h\ 48^m\ 31^s.1$ plus $18^h\ 02^m\ 57^s.4$, or (after subtracting 24 hours) $7^h\ 51^m\ 28^s.5$.

The following form of solution may be used:

1915			
LMT, 91:31:30 W, civil, October 20...... A.M.			$6^h\ 00^m\ 00^s.0$
LMT, 91:31:30 W, astronomical, October 19..			18 00 00 .0
Correction, solar to sidereal.................		+	2 57 .4
Sidereal interval since mean noon...........			$18^h\ 02^m\ 57^s.4$
A, at Greenwich mean noon, October 19.................	13 47 30.95		
Increase due to longitude.......	1 00.15		
A_n...................................			13 48 31 .1
Sidereal time on October 19, solar (astr.) date.			$31^h\ 51^m\ 28^s.5$
or			7 51 28 .5

31. To Change Sidereal to Local Mean Solar Time.

Problem: What was the local mean time on October 19, 1915, (astronomical date) in longitude 91° 31′ 30″ W when the sidereal time was $7^h\ 30^m\ 27^s.5$?

Read Art. 21, page 23.

Solution: The solution is much the same—except that the order is reversed—as that of the preceding problem, the increase in the right ascension of the mean sun being taken care of in the same way.

From the given sidereal time subtract A_n (first adding 24 hours to the sidereal time if necessary for the subtraction). The difference is the sidereal interval since local mean noon. Change this sidereal interval to the corresponding solar interval by use of Table I (or Table II of the "Nautical Almanac") and the result is local mean time.

Stated in other words, the result of the subtraction is the number of hours of arc which the vernal equinox has covered since local mean noon. From this result is computed the number of hours of arc which the mean sun has covered since local mean noon; *i.e.*, the local mean time.

It should be clearly understood that adding 24 hours to, or subtracting 24 hours from, the hour angle of the vernal equinox does not in any way affect the date; for the date bears no relation to the hour angle of the vernal equinox—*i.e.*, to the sidereal time—but is dependent only on the sun. In other words, dates are always solar dates.

The following form of solution may be used for the problem above:

	1915	
Sidereal time, 91:31:30 W, Oct. 19, astr. date..		$7^h\ 30^m\ 27^s.5$
A, at Greenwich mean noon, Oct. 19......................	13 47 30.95	
Increase due to longitude.......	1 00.15	
A_n....................................		13 48 31 .1
Sidereal interval since mean noon............		$17^h\ 41^m\ 56^s.4$
Correction, sidereal to solar.................	−	2 54 .0
LMT, astronomical, 91:31:30 W, Oct. 19.....		$17^h\ 39^m\ 02^s.4$
LMT, civil, 91:31:30 W, Oct. 20 ........A.M.		5 39 02 .4

CHAPTER VI

OBSERVATIONS—CORRECTIONS TO OBSERVATIONS

32. Objects Observed. Methods of Naming Stars. Observations are to be made on the sun and on the stars. The moon and the planets are sometimes used as objects for observations, especially for longitude; but these observations more properly form a part of geodetic work of greater refinement than is herein contemplated, and their discussion will be omitted.

The stars are distinguished according to the following scheme: The sky is divided into irregular areas, usually such that the stars in a given division seem to form a natural group; and all the stars within that area form a constellation, which receives a name. The individual stars of a constellation are sometimes distinguished by receiving a special name, and usually by a Greek letter or a number also. The letters of the Greek alphabet are usually assigned to the stars of a constellation in descending order of brightness; α to the brightest, β to the next, and so on. A star is then named by stating its letter followed by the name of the constellation to which it belongs in the Latin genitive form. Thus the "pole-star" has the special name "Polaris," and since it is the brightest star in the constellation "Ursa Minor," it is also called "α Ursæ Minoris." Sometimes two stars which are apparently very close together are given the same letter; in which case a small number is placed over their letter, as α^1, α^2, etc., to distinguish them and to indicate the order in which they cross the meridian.

The brightness of a star is designated by a number from a numerical scale of magnitudes. In this scale the numbers increase as the magnitudes decrease. Stars of the fifth magnitude are about as dim as can be seen under favorable conditions with the naked eye. Polaris is of the second magnitude.

33. Circumpolar Constellations. Stars whose co-declination or polar distance is less than the latitude of the observer do not go below his horizon at any point in their diurnal circles, and are called circumpolar stars. Those circumpolar stars which are near the pole are of most importance to the surveyor.

Some of the more important of these are here shown in Fig. 12.

The one most used of all because it is the nearest the pole is the brightest star in the constellation "Ursa Minor," or the "Little Dipper." It is Polaris, or α Ursæ Minoris. It is about 1° 08′ from the pole (1916) and is approaching the pole at the rate of about 0′.3 per year. There is no star exactly at the pole. No other bright star is near Polaris which is likely to be confused with it; and it is easy to find by reference to two stars in the constellation Ursa Major, or the "Great Dipper," on the opposite side of the pole. The two brightest stars in this constellation, the ones which form the side of the "bowl" opposite the "handle" of the Dipper, are called the "Pointers." A line drawn through them and produced falls very near Polaris.

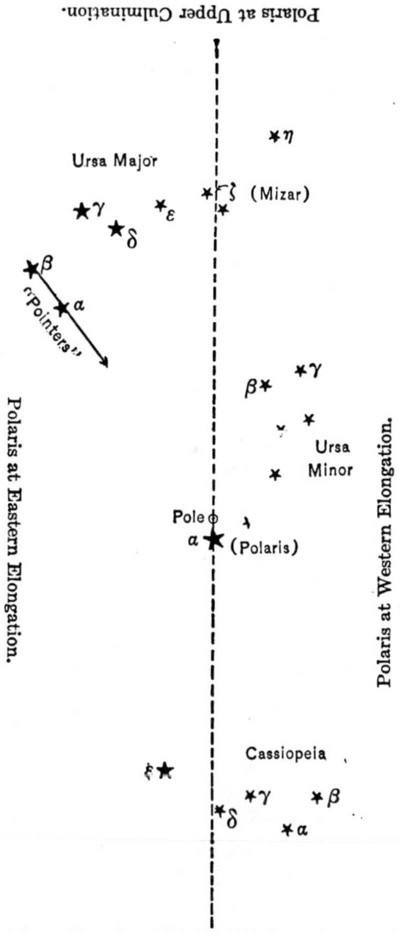

FIG. 12.

SOME OF THE CIRCUMPOLAR CONSTELLATIONS
(At the North Celestial Pole.)

The star ζ Ursæ Majoris, at the bend in the handle of the Great Dipper (see Fig. 12), is of some use to the surveyor because it falls very nearly on the same hour circle as Polaris and δ Cassiopeiæ. Cassiopeia is a constellation on the opposite side of Polaris from Ursa Major, the five brightest stars of which form a rather awkward "*W*". δ Cassiopeiæ is at the lower left-hand corner of the *W*. The position of Polaris in its diurnal path around the pole may be estimated quite accurately by the relative positions of these three stars: Polaris, Ursæ Majoris, and δ Cassiopeiæ. If they are in a vertical line with δ Cassiopeiæ above, Polaris is at upper culmination; if ζ Ursæ Majoris is above, Polaris is at lower culmination. If they are in a horizontal position with δ Cassiopeiæ at the right, or east, Polaris is at eastern elongation; while a reversed position indicates western elongation of Polaris.

β Cassiopeiæ, at the upper right-hand corner of the *W*, has a right ascension very nearly equal to zero; *i.e.*, it is on an hour circle which passes very near the vernal equinox. Therefore, the hour angle of this star is closely equal to local sidereal time. This hour angle, and thereby local sidereal time, may be estimated fairly well by remembering that when the star is vertically above Polaris it is practically 0 hours, when it is vertically below it is 12 hours; and that the points half-way between these two positions on the left and right correspond to 6 hours and 18 hours, respectively.

In the determination of latitude and of time we shall find use for some of the stars in other constellations than those mentioned, but we can identify them as needed by means of their co-ordinates (taken from the "American Ephemeris and Nautical Almanac," in terms of right ascension and declination, and converted into other co-ordinates for use) and they will not be discussed further at present.

34. Parallax. The co-ordinates of a celestial object should be referred to the center of the celestial sphere as the pole or origin. This center is at the center of the earth; and an altitude of any body nearer than the fixed stars—such as the sun—which has been measured from the surface of the earth, is less than the altitude referred to the center of the earth as the origin of co-ordinates by an amount called the **parallax,** and must be reduced to the value for the center by applying a **parallax correction.** If the earth is assumed to be a sphere, which is sufficiently accurate for practical purposes, the effect of parallax is to decrease the altitude of a body without affecting the azimuth.

Referring to Fig. 13, the angle $H'AS$ is the measured altitude of a point, S, obtained by an observation from A, a point on the surface of the earth; and HOS is its altitude as referred to the center of the earth, O. The difference between the two, or the angle ASO, is the parallax correction.

In the triangle AOS,

$$\sin ASO = \sin OAS \frac{OA}{OS} \quad . \quad . \quad . \quad . \qquad \text{(a)}$$

where the angle OAS is equal to 90° plus the measured altitude,

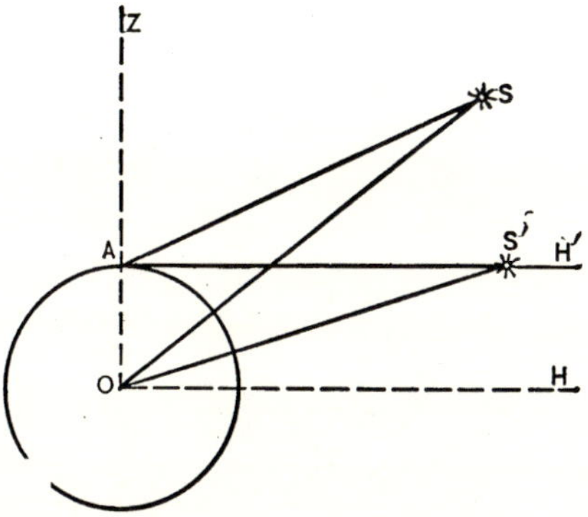

FIG. 13

OA is the radius of the earth, OS is the distance from the center of the earth to the center of the body observed.

When the observed body is at the zenith it is evident that the parallax will be zero; and when it is on the horizon, as observed from the surface of the earth (*i.e.*, at S'), its parallax will be a maximum. This maximum is called the **horizontal parallax.**

For this case, where angle OAS' is equal to 90°:

$$\sin AS'O = \frac{OA}{OS'} \quad . \quad . \quad . \quad . \quad . \quad . \qquad \text{(b)}$$

Let: c_p = parallax at any position of the body.
C_p = horizontal parallax.
h' = measured altitude.

Since $OS = OS'$, Equation (b) may be written:

$$\sin C_p = \frac{OA}{OS}.$$

Substituting this value for OA/OS in Equation (a), and remembering that $\sin OAS = \cos h'$, we obtain:

$$\sin c_p = \sin C_p \,.\, \cos h' \quad . \quad . \quad . \quad . \qquad \text{(c)}$$

Since c_p and C_p are very small angles we may without appreciable error substitute the angles for their sines, and Equation (c) then becomes:

$$\mathbf{c''_p = C''_p \,.\, \cos h'} \quad . \quad . \quad . \quad . \quad . \qquad (21)$$

in which c_p and C_p are both expressed in seconds of arc.

For the sun the mean value of Cp is 8″.8. For the moon and for the planets it is much larger. For the stars it is too

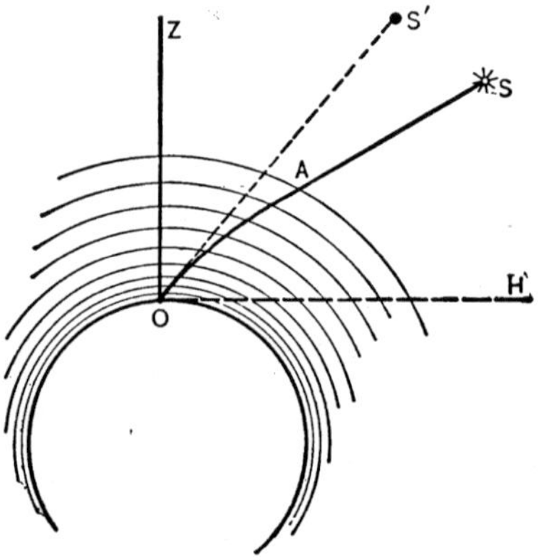

FIG. 14.

small to be measured, because of their great distance from the earth. There is, therefore, no parallax correction to measured altitudes of the fixed stars.

The parallax correction, when needed, is always to be added to observed altitudes.

Parallax corrections to measured altitudes of the sun are given in Table IV.

35. Refraction. When a ray of light passes from one medium to another of different density it is bent, and this bending is called **refraction.** As a ray of light comes from a celestial body to the eye of an observer it passes through successive layers of atmosphere of increasing density, and is therefore bent downward in a curve, as shown in Fig. 14.

If an observer on the surface of the earth, at O, is looking at a star which is actually at S, it will appear to him to be at some point above S, as at S'; OS' being a tangent at O to the curve AO. The angle which must be subtracted from the measured altitude, HOS', to obtain the altitude of S is called the **refraction correction.**

A convenient rule for the amount of this correction, derived by application of the laws of physics after some simplifying assumptions have been made, is that the refraction correction in minutes is equal to the natural co-tangent of the observed altitude. This is sufficiently accurate for all altitudes greater than about ten degrees which have been measured with ordinary field instruments—transit or sextant. Differences in temperature and in barometric conditions cause differences in the amount of the refraction correction, but they are too slight to require attention in work of this character.

Table III gives the amount of the refraction correction to be applied to observed altitudes for a mean barometric pressure and temperature.

The refraction correction is always to be subtracted from an observed altitude.

36. Semi-diameter. The sun's disc, as seen through a telescope, is circular, and measurements are usually made to its edge, or limb, rather than to its center. Such measurements must be corrected by the amount of the angle subtended by the semi-diameter of the sun to reduce them to the proper values for the center. As has been mentioned, the amount of this angular semi-diameter is given for each day in the year in the "Nautical Almanac"; and may be taken therefrom for use in reducing observations.

Values of the sun's semi-diameter for the first of each month in the year are given in Table IV at the back of this book. Interpolations from this table are sufficiently accurate for most work with field instruments.

37. Instrumental Errors. Though it would be well if an instrument in perfect adjustment could be used for all astronomical observations, it is not always possible; and care should be taken to eliminate, so far as may be, the effect of any inaccuracy of adjustment or construction.

Considering first the transit: When it is in use care should be taken to make the plates truly horizontal. If the plate bubbles are not in good adjustment this can still be accomplished by leveling up, turning the plates 180°, and by means of the leveling screws bringing each bubble half-way back to the center of its tube. When the plates are level both bubbles should remain in the same positions in the tubes throughout a complete revolution of the plates.

Whenever possible horizontal angles should be repeated—at least doubled—making half of the measurements with the telescope direct and half with it inverted. The average of the results is then free from the effect of errors in the line of sight or in the height of standards, and should be of greater precision than any one reading.

Vertical angles cannot be repeated; but whenever time permits at least two readings should be taken, one with the telescope direct and one with it inverted, releveling if necessary after reversing the instrument. This, of course, presupposes the use of a transit with a full vertical circle. The mean of the two readings should be free from the effect of errors in adjustment in line of sight, standards, telescope level, and index error. If, because the instrument does not have a full vertical circle or because of lack of time, the above method cannot be used, care must be taken to see that the axis of the telescope level is parallel to the line of sight; and the index error must be determined and proper correction made. The index error is the reading of the vertical arc or circle when the line of sight is horizontal.

It may be convenient to remember that for use with vertical angles read *above* the horizontal, if the zero of the vernier is to the right of the zero of the vertical arc when the telescope is horizontal the index correction is positive; while if the zero of the vernier is to the left of the zero of the vertical arc when the telescope is horizontal the index correction is negative.

Concerning the sextant: The sextant is not adapted to the measurement of horizontal angles between objects at different elevations; but vertical angles may often be measured with greater precision with the sextant than with the transit, on account of the greater radius and finer graduation of the limb. There is usually an index error, whose amount should be determined and the readings properly corrected.

The amount and sign of the index error may be determined in the following manner: Using an artificial horizon, bring the direct and reflected images of the sun externally tangent to each other in each of the two possible positions, and read the vernier at each setting. It will be noticed that it is necessary to consider the numbering of the divisions on the vernier reversed when making the reading to the right of the zero of the main limb. Call the reading to the left minus and that to the right plus. Half the algebraic sum of the two readings is then the index error, with the proper algebraic sign. It should be remembered that the index *correction* will have the opposite sign.

38. Sequence of Corrections. Corrections to observed altitudes should be made in the following order:

(1) Instrumental corrections.
(2) Refraction correction.
(3) Semi-diameter correction.
(4) Parallax correction.

The chief and probably the only instrumental correction that can be applied if ordinary field instruments are used will be the index correction. Care should be taken to give it the proper sign.

The algebraic sum of the refraction and parallax corrections is often applied as a single correction.

When applying a refraction correction to an observed altitude of the sun, the correction for the limb observed—not for the center—should be used. Because they are at different altitudes the values of the corresponding corrections would differ considerably if the altitudes were small.

39. Suggestions for Observing. When making astronomical observations greater care is necessary in the instrumental work than is often exercised in ordinary surveying.

Considering first the use of the transit:

A good, firm "set-up" should be secured; and the two plates of the leveling head should be nearly parallel when the instrument is leveled. The leveling screws should be turned to just the proper degree of tightness—not too loose so as to allow the instrument to rock, nor so tight as to bind and later spring the plates from a horizontal position. So far as possible the effect of instrumental errors should be eliminated by the method of observing.

Extra care is needed in reading angles at night; that is, in determining the reading of the vernier at a given setting. A lantern—or better, an electric flash-light—should be held beside and rather back of the head of the observer. It will be noticed that there is greater likelihood of getting a wrong reading because of not looking squarely down on the vernier than by daylight.

When observing at night the cross-hairs must usually be

illuminated in order to be seen. A reflector made by an instrument-maker is sometimes used. It is simply a cylinder to be put on the telescope in place of the sunshade, from the side of which a large "notch" has been cut. This notch is framed with brightly polished metal, so placed as to reflect light down the barrel of the telescope from a lantern held beside it.

If such a reflector is not at hand one which will serve equally well may be made from a piece of white paper three or four inches wide and long enough to wrap around the objective end of the telescope, where it is held in place by a rubber band. A crescent-shaped cut in the side of the paper cylinder thus formed will produce a flap which may be pushed inward to act as the reflector.

Before attempting to "find" a star with the telescope the eye-piece should be properly focused on the cross-hairs and the objective focused on a distant object, such as a distant light, and then they should not be disturbed; as it is difficult to find a star if the telescope is not properly focused. This focusing, especially that of the eye-piece, may profitably be done before dark. The star may be found by sighting first along the telescope and then through it before a light is brought near the instrument. When the star is clearly visible as a point of light the lantern should be gradually brought nearer the reflector until the cross-hairs may be seen distinctly, but not so close as to make the field so bright that the star cannot be plainly seen. The star will not appear any larger or brighter through the telescope than when seen by the naked eye.

Suggestions in regard to "lining in" a point at night, or establishing a reference-mark from which horizontal angles may be measured at night, are given in connection with the work in which these operations are required; namely, at the beginning of Chapter VIII on "Observations for Azimuth." The methods of lining in a stake and tack serve equally well to suggest methods of sighting on a stake already set.

When making observations on the sun the eye-piece of the telescope must be covered with a piece of dark glass to protect the eye. For measuring altitudes greater than fifty or sixty degrees a prismatic eye-piece must be used, fastened on over the regular eye-piece. When using this attachment in making observations on the sun it should be remembered that the prism turns the image upside down, but not right for left. If either the colored glass or the prismatic eye-piece is not at hand when needed the following scheme may be used, provided the telescope has an erecting eye-piece:

Draw the eye-piece nearly out and the objective nearly in by means of the focusing screws. If now the telescope is pointed toward the sun and a white card held below and three or four inches from the eye-piece, the images of the sun and of the cross-hairs may both be focused on the card by slight manipulations of the focusing screws; and a pointing may be made quite accurately without actually looking through the telescope. If the eye-piece is non-erecting this method cannot be used.

Concerning the use of the sextant:

For the theory, adjustments, and general method of use of the sextant the reader is referred to any standard text on surveying.* The method of determining the index error has been given in Art. 37, page 42.

As has been noted, the sextant is not adapted to the measurement of horizontal angles between points not at the same elevation; so that its use in the work described in this book will be limited to the observation of the altitude of the sun or of a star. Measuring the altitude of a celestial body with a sextant consists in measuring the angle between the object itself and its reflection from a surface called an "artificial horizon."

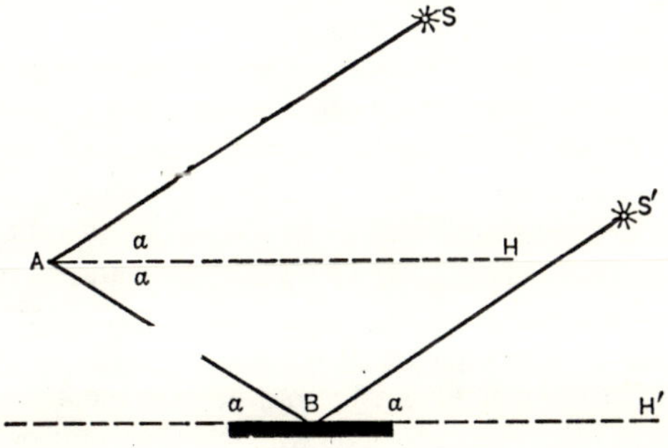

FIG. 15.

In Fig. 15, A represents the position of the eye of the observer, B the artificial horizon, and SA and $S'B$ rays from the same celestial object to the eye and to the artificial horizon, respectively. $S'B$ is reflected along the line BA. AH is a horizontal

* See, for instance, Raymond's "Plane Surveying," Second Edition, page 393; or Breed and Hosmer, "Principles and Practice of Surveying," Volume II, page 274.

line through the eye of the observer, and BH' is a horizontal line as determined by the surface of the artificial horizon. It is evident that all the angles marked α in the figure are equal; and that the measured angle, SAB, is therefore equal to twice the apparent altitude of the body S. The angle is measured by observing through the transparent portion of the horizon glass of the sextant the image reflected from the artificial horizon, and bringing into coincidence (or tangency) with it the image of the object as reflected from the index glass.

When measuring the altitude of the sun, using either no telescope or an erecting telescope, if the apparent lower limb of the sun as reflected from the index glass is brought into contact with the apparent upper image seen in the artificial horizon, the angle measured is twice the altitude of the sun's lower limb. If the telescope is an inverting one, the angle measured by this method is twice the altitude of the upper limb. The index correction should be applied before the measured angle is divided by two to obtain the altitude.

A shallow dish of mercury is usually considered most satisfactory for an artificial horizon, but a dish of molasses will answer nearly or quite as well. Whatever is used, it should be protected from disturbance from the wind or other causes during the observations. A roof-shaped cover with glass windows is usually provided for use with the mercury horizons. This is satisfactory if the two faces of each piece of glass are parallel planes. A cover of fine mosquito netting will serve the purpose quite well, and introduce no error from refraction.

A good deal of care and considerable practice are required to obtain accurate results from the use of the sextant; but because of its finer graduation, which is made possible on account of the greater radius of the limb as compared with a transit (sextants are commonly graduated to read to the nearest ten seconds, or at least to the nearest half minute), it is capable of giving very precise results. Because of its portability it is adapted for use in places where a transit could not be used. Since sights to both objects which determine the angle to be measured are taken at the same time it is not necessary that the sextant have a firm support, such as is required for the transit. It is the instrument used for astronomical observations at sea; in which case the sea horizon instead of an artificial horizon is used from which to measure altitudes, requiring a correction for the "dip" of the apparent below the true horizon.

There is usually furnished with the sextant a special telescope for astronomical work. There are several colored glasses attached to the frame which may be turned into the line of sight to protect the eye when the sun is the object observed. It is a good plan to use one color in front of the horizon glass and a different color in front of the index glass, so that the two images may be of different color and be more easily distinguished.

One final suggestion which applies equally well, no matter what instrument is used or what the observation may be, is:

Before going into the field the observer should have clearly and definitely in mind exactly what things he is to do, and exactly how and when and in what order he is to do them. This is important.

CHAPTER VII

OBSERVATIONS FOR LATITUDE

40. Latitude by a Circumpolar Star at Culmination. This observation consists in measuring the altitude of a circumpolar star at upper or lower culmination, when its altitude is a maximum or a minimum, and from this measured altitude and known data computing the altitude of the north celestial pole, which is the latitude of the place. Any circumpolar star may be used, but Polaris is the best, because it is the brightest.

Referring to Fig. 16, let the circle represent the meridian of the observer, who is at O. Let HH' and EE' represent the projections of the horizon and celestial equator, respectively, upon the plane of the meridian. Let Z represent the zenith and P the north celestial pole.

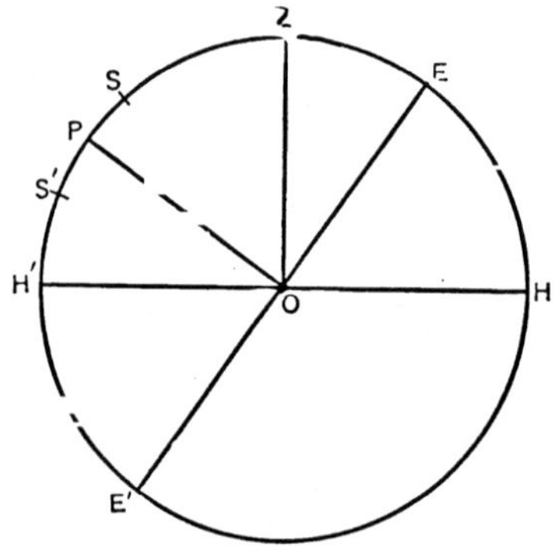

FIG. 16.

By definition, the arc EZ (the angular distance of the observer from the equator) is the observer's latitude. It is obvious that the arc $H'P$ is equal to the arc EZ, and that therefore the observer's latitude may be defined as the declination of his zenith (EZ) or as the altitude, with respect to his horizon, of the celestial pole ($H'P$).

Let S and S' be the two positions of a circumpolar star when it is on the meridian of the observer, at upper and lower culmination, respectively. Considering first the case of upper culmination, the declination of the star is ES, and its altitude is $H'S$. Also:

$$H'P = H'S - PS$$
$$= H'S - (90° - ES)$$

or

$$\phi = \mathbf{h} - (90° - \delta) \quad . \quad . \quad . \quad . \quad . \quad (28)$$

For the case of lower culmination, when the star is at S', the declination is $E'S'$, the altitude is $H'S'$, and:

$$H'P = H'S' + PS'$$

or

$$\phi = \mathbf{h} + (90^\circ - \delta) \quad . \quad . \quad . \quad . \quad . \quad (29)$$

If Polaris is the star used it is not strictly necessary that the exact time of culmination be known, for the altitude of Polaris changes but very slightly for several minutes before and after culmination. The time of culmination may be taken from Table V (in which case the declination may be taken from Table VI), or it may be computed more exactly by the following method. This method applies equally well to any circumpolar star.

At the instant of upper culmination the hour angle, t, of any star is equal to zero, and at the instant of lower culmination it is equal to 12 hours. The right ascension of the star for any desired date may be found in the "American Ephemeris and Nautical Almanac." We may therefore compute the sidereal time of culmination by the following equation:

$$\textbf{Sid. T} = \textbf{R A} + \textbf{t} \quad . \quad . \quad . \quad . \quad . \quad (24)$$

This sidereal time may be changed to standard time by the methods of Arts. 31 and 27, pages 34 and 31; thus giving the standard or watch time of culmination.

The longitude of the place may be obtained with sufficient accuracy for this or any similar computation by scaling from one of the government's topographical sheets or any other reliable map, even if drawn to a very small scale. An error of half a degree in longitude represents only about two minutes' error in computed time of culmination, and the altitudes of any of the close circumpolar stars change but slightly for several minutes before and after culmination.

If no tables or means of computation are at hand the approximate time of upper or lower culmination of Polaris may be estimated by the relative positions of Polaris and δ Cassiopeiæ. See Art. 33, page 38, and Fig. 12. Beginning some little time before culmination, the motion of Polaris may be followed by the tangent screw of the vertical motion of the transit, bisecting the star with the horizontal cross-hair, until it has reached its highest or lowest position and appears to have only a horizontal motion. Its altitude should then be read.

It is, of course, unnecessary that the instrument be centered over any definite station during this observation, as a difference of a minute in latitude corresponds to about 6080 feet on the ground. Either an engineer's transit or a sextant and artificial horizon may be used in this observation if the time of culmination has been computed.

Outline of Observation:

Computations Preceding Field Work:

Compute time of U. C. or of L. C. of star.

From Table V (for Polaris only), *or*, more accurately,

$$\textbf{Sid. T} = \textbf{R A} + \textbf{t} \quad . \quad . \quad . \quad . \quad . \quad (24)$$

R A from "Nautical Almanac."

$t = 0^h$ for U. C., $t = 12^h$ for L. C.

Change *Sid. T* to *Std. T*.

Field Work:

(A) Using a sextant:

Make several measurements of double altitude of star within three minutes of time of culmination

Determine index error of sextant.

(B) Using a transit with vertical arc:

Beginning several minutes before computed time of culmination, follow star with tangent screw to limit of star's vertical motion, bisecting it with horizontal cross-hair.

Read the vertical arc, and determine index error.

(C) Using a transit with vertical circle:

With telescope direct, read altitude of star two or three minutes before culmination.

Reverse instrument quickly and read altitude of star with telescope inverted.

Computations Following Field Work:

(A) Apply index correction to measured angle.

Divide result by two for apparent altitude.

Subtract refraction correction from apparent altitude, thus obtaining true altitude, *h*.

(B) Correct vertical arc reading for index error and refraction, thus obtaining true altitude, *h*.

(C) Subtract refraction correction from mean of two readings, thus obtaining true altitude, *h*.

Having true altitude, *h*, apply equation:

$$\phi = \mathbf{h} \mp (90° - \delta) \quad . \quad . \quad . \quad . \quad . \quad . \quad . \quad (28), (29)$$

Use − sign for U. C.
Use + sign for L. C.
Obtain δ from Table VI (for Polaris only), *or*, more accurately from the "Nautical Almanac."

An example of the computations and field-notes of this observation is given on pages 106 and 107.

41. Latitude by Meridian Altitude of a Southern Star. This observation consists in measuring the altitude of a southern star when it is on the observer's meridian—*i.e.*, at upper transit—and from this measured altitude and known data computing the declination of the observer's zenith, which is his latitude.

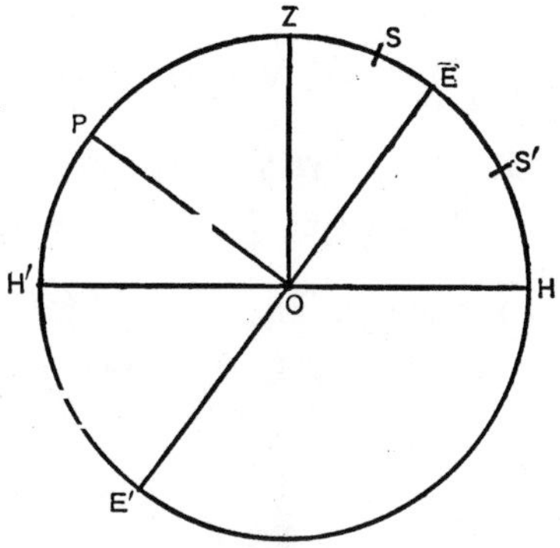

FIG. 17.

Referring to Fig. 17, a southern star may cross the meridian between the equator and the zenith, as at S, or below the equator, as at S'. The lower transit of a southern star, assuming the observer to be in the northern hemisphere, is always invisible.

Considering the first case, the declination, ES, is positive; and

$$EZ = 90° - (HS - ES),$$

or

$$\phi = \mathbf{90°} - (\mathrm{h} - \delta) \quad . \quad . \quad . \quad . \quad . \quad (30)$$

If the star is at S' the declination, ES', is negative, and the equation may be written:

$$EZ = 90° - HS' - (-ES')$$

That is, if δ is always substituted with its proper algebraic sign the equation derived above is general, *viz.:*

$$\phi = \mathbf{90°} - (\mathbf{h} - \delta) \quad . \quad . \quad . \quad . \quad . \quad (30)$$

The sidereal time at which the star will cross the meridian may be computed from the equation:

$$\textbf{Sid. T} = \textbf{R A} + \textbf{t} \quad . \quad . \quad . \quad . \quad . \quad (24)$$

the right ascension of the star for the proper date being taken from the "American Ephemeris and Nautical Almanac," and the hour angle, t, for the instant being equal to 0 hours. The sidereal time thus computed may be changed to standard time by the methods of Arts. 31 and 27, pages 34 and 31.

One objection to this observation is the difficulty in identifying the southern stars. If the direction of the meridian and the approximate latitude are known, the methods of identification of Art. 48, page 68, may be used.

Outline of Observation:

Computations Preceding Field Work:

Compute time of transit of star.

$$\text{Sid. T} = \text{R A} + \text{t} \qquad (24)$$

R A from " Nautical Almanac."

$t = 0$ hours.

Change *Sid. T* to *Std. T*.

Field Work:

(A) Using a sextant:

Make several measurements of double altitude of star within two or three minutes of time of transit.

Determine index error of sextant.

(B) Using a transit with vertical arc:

Beginning several minutes before time of transit, follow star with tangent screw to limit of star's vertical motion, bisecting it with horizontal cross-hair.

Read the vertical arc and determine index error.

(C) Using a transit with vertical circle:

With telescope direct, read altitude of star two or three minutes before transit.

Reverse instrument quickly and read altitude of star with telescope inverted.

Computations Following Field Work:

(A) Apply index correction to measured angle.

Divide result by two for apparent altitude.

Subtract refraction correction from apparent altitude, thus obtaining true altitude, h.

(B) Correct vertical arc reading for index error and refraction, thus obtaining true altitude, h.

(C) Subtract refraction correction from mean of two readings, thus obtaining true altitude, h.

Having true altitude, h, apply equation:

$$\phi = 90^\circ - (h - \delta) \quad . \quad . \quad . \quad . \quad . \quad . \quad (30)$$

Obtain δ from "Nautical Almanac," and substitute it with proper algebraic sign.

An example of the computations and field-notes of this observation is given on pages 108 and 109.

42. Latitude by Meridian Altitude of the Sun. The methods and equations of the preceding article may be applied to the sun as well as to a southern star. The altitude of one edge, or limb, of the sun instead of the altitude of the center is usually measured. (See Arts. 36 and 38, pages 41 and 43.)

The watch (standard) time of transit of the sun's center may be obtained by changing 0 hours local apparent time to standard time by the method of Art. 29, page 32. If the direction of the meridian is known, the meridian altitude may be obtained by use of a transit without computation of the time of transit of the sun, by setting the instrument so that the telescope revolves in the plane of the meridian and reading the altitude of one limb of the sun when its center crosses the vertical cross-hair. To make the telescope revolve in the plane of the meridian, set up the transit over one of two stakes which mark the direction of the meridian and sight on the other. The line of sight will now revolve (about the horizontal axis of the telescope) in the plane of the meridian. In order to eliminate the effect of instrumental errors so far as possible, it is well to set over the farther north of the two stakes and sight on the south stake.

The declination of the sun at the instant at which the observation is to be made (local apparent noon) may be obtained, by interpolation, from the "Nautical Almanac" or from one of the instrument-makers' reprints from the "Nautical Almanac." The longitude, for use in making the interpolation, may be obtained by scaling from some reliable map, such as one of the government's topographical sheets. This value of the longitude should also be sufficiently accurate for use in the conversion of time mentioned above.

Outline of Observation:

Computations Preceding Field Work:

Compute the standard time of transit of the sun's center.

Change 0 hours, local apparent time, to standard time.

Obtain longitude by scaling from a map, other data from "Nautical Almanac."

Field Work:

(A) Using a sextant:

Measure double altitude of lower limb of sun at computed time of transit. (See Art. 39, page 43, for method of using sextant.) Determine index error.

(B) Using a transit:

Follow the sun with tangent screw of vertical motion as long as it rises, keeping the horizontal cross-hair tangent to the sun's lower limb. (See Art. 39, page 43, for suggestions in regard to sighting on the sun.)

Take the reading of the vertical arc (or circle) corresponding to the greatest altitude of the sun. (Should occur practically at computed time of transit.)

Determine index error.

Computations Following Field Work:

(A) Apply index correction to measured angle.

Divide result by two for apparent altitude of lower limb.

Apply refraction, semi-diameter, and parallax corrections, thus obtaining true altitude, h.

(B) Correct vertical arc (or circle) reading for index error, refraction, semi-diameter, and parallax, thus obtaining true altitude, h.

Having true altitude, h, of the sun's center, apply equation:

$$\phi = 90° - (h - \delta) \quad . \quad . \quad . \quad . \quad . \quad . \quad (30)$$

Obtain δ for the time of observation from the "Nautical Almanac" (using longitude scaled from a map for interpolation), and substitute it with proper algebraic sign.

An example of the computations and field-notes of this observation is given on pages 110 and 111.

CHAPTER VIII

OBSERVATIONS FOR AZIMUTH

For the engineer in general practice the determination of true azimuth is probably the most important part of the work of field astronomy.

In the discussion of all observations for azimuth it will be assumed that the transit is carefully set up and centered over a point which marks one end of a line whose azimuth is desired. This position of the instrument will be called simply the "station." In some observations it will be most convenient to line in another "point"—a stake and tack—in the direction of a star which has been sighted, and whose azimuth at the instant of sighting can be computed; in other observations it will be best to measure the angle from the star or the sun to a signal. In any case, the true azimuth of a line on the ground, one end of which is the station and the other a stake or signal, is obtained. From this reference line we may then determine the azimuth of any other line which passes through the station, or we may determine the direction of the meridian through the station. It is customary to call the angle between the true north and the direction of a circumpolar star the azimuth of the star, without regard to whether the star is east or west of the true north. This azimuth is the angle Z of the astronomical triangle.

If a stake is to be lined in at night it may be conveniently done in the following manner:

First line in a lantern. Then hold in front of the lantern an oiled paper screen, and in front of the screen a stake. (A suitable screen may be made by tacking some heavy paper to four sticks nailed together to form a rectangular frame, and pouring some kerosene or other oil over the paper; or a handkerchief will do fairly well in place of a screen.) The stake can then be seen, black against the bright screen, and lined in in the ordinary manner. The cross-hairs of the transit can often be seen against the screen, thus avoiding the necessity of otherwise illuminating them. The stake should be driven where a pencil held on the top can be seen, to be lined in for the exact point. The screen is not necessary, but it furnishes a larger, more uniformly lighted area against which to see the stake than does the lantern alone.

A suitable signal, or "azimuth mark," from which to measure an angle to a star, may be made from a wooden box large enough to hold a lantern. A hole should be bored in the front of the box at the height of the blaze of the lantern, the size of the hole depending on the distance from the station at which the box is to be set, and on the strength of the light behind it. These factors should be so adjusted that the appearance at night will be that of a point of light not unlike a star. If an ordinary kerosene lantern is used, a hole half an inch in diameter in a box a quarter of a mile or so away will give a suitable mark to sight on. The distance from the station should be great enough so that the focus of the telescope will not have to be changed when sighting first at the signal and then at a star. A vertical line through the hole which can be used as a sight from the station in the daytime should be painted on the box. The box should be covered to prevent the lantern being blown out; and it should be nailed to a tree or to stakes driven firmly in the ground approximately north of the station, and where an unobstructed view of it may be had from the instrument.

It is a good plan to line in a stake between the station and the mark and near the latter, so that its direction will not be lost if the box is destroyed.

43. Azimuth by a Circumpolar Star at Elongation. This method is probably under ordinary conditions the most reliable means of determining true azimuth.

The observation consists in measuring the angle from a circumpolar star at elongation (eastern or western) to an azimuth mark, and from the measured angle and the computed azimuth of the star at the instant of elongation computing the azimuth of the mark from the station. If desired, instead of measuring the angle to an azimuth mark, a stake and tack may be lined in in the direction of the star. Any circumpolar star may be used, but Polaris is the best. If Polaris is used the time of elongation may be taken from Table V; or it may be more accurately computed by the following method, which applies to any circumpolar star.

The sidereal time of elongation may be obtained from the equation:

$$\text{Sid. T} = \text{R A} + t \quad . \quad . \quad . \quad . \quad . \quad (24)$$

The right ascension for the proper date may be taken from the "Nautical Almanac," and the value of t found by a solution of the astronomical triangle. See Art. 12, Solution (2), page 14, and Art. 9, pages 9 to 12. The sidereal time thus obtained may be changed to standard time by the methods of Arts. 31 and 27, pages 34 and 31.

If no tables or means of computation are at hand, the approximate time of elongation of Polaris may be estimated by the relative positions of Polaris and δ Cassiopeiæ (see Art. 33, page 38, and Fig. 13) and the motion of Polaris followed with the tangent screw of one of the horizontal motions of the transit (bisecting the star with the vertical cross-hair) until it ceases to move east or west as the case may be, and appears to be moving vertically.

The azimuth of Polaris when at elongation may be taken from Table VII, or it may be more accurately computed by Formula (12), Art. 12, page 15; the declination being taken from the "Nautical Almanac," for the proper date and the latitude being known from a previous observation or from an accurate map.

Outline of Observation:

Computations Preceding Field Work:

Compute time of elongation of star—eastern or western.

From Table V (for Polaris only), *or* more accurately,

$$\text{Sid. T} = \text{R A} + \text{t} \quad \text{. . . . .} \quad (24)$$

R A from "Nautical Almanac."

$t = P$ (western elongation) *or*

$t = 24^h - P$ (eastern elongation).

$$\cos \text{P} = \frac{\tan \phi}{\tan \delta} \quad \text{. . . . .} \quad (13)$$

ϕ from a previous observation or from a map .

δ from "Nautical Almanac."

Change *Sid. T* to *Std. T*.

Field Work:

Sight on azimuth mark with plates set at zero.

Turn to star with upper motion of transit and follow star with upper tangent screw, bisecting star with vertical cross-hair.

Three or four minutes before time of elongation, when star appears to move vertically, read the horizontal angle between mark and star.

Double the angle.

Field Work *(Continued):*

Reverse the instrument quickly and double the angle with telescope inverted.

Computations Following Field Work:

Compute azimuth of star at elongation.

From Table VII (for Polaris only), *or* more accurately,

$$\sin Z = \frac{\cos \delta}{\cos \phi} \quad . \quad . \quad . \quad . \quad . \quad . \quad . \quad . \qquad (12)$$

δ from " Nautical Almanac."

ϕ from a previous observation or from a map.

From computed azimuth of star and mean of measured angles between star and azimuth mark, compute the azimuth of the mark.

Or, if desired, the field work may be done as follows:

Field Work:

Sight on star and follow it with tangent screw of either horizontal motion of transit until two or three minutes before elongation, when star appears to move vertically.

Plunge telescope down and line in a stake and tack several hundred feet away.

Reverse instrument quickly, sight on star with telescope inverted, and line in another point beside the first.

The mean of these two points should be in the direction of the star at elongation.

Computations as outlined above.

An example of the computations and field-notes of this observation is given on pages 112 and 113.

44. Azimuth by Polaris Near Elongation. If the observation described in the preceding article is made on *Polaris* within thirty minutes of elongation, the azimuth of the star at the instant of each sight may be obtained from the computed azimuth at elongation by the following formula:

$$\mathbf{C = 3600 \times 112.5 \times \sin 1'' \times \tan Z_e \times T^2} \quad . \quad (31)^*$$

where T is the interval in (sidereal) minutes between the instant of elongation and the instant of sighting.

---

* The demonstration of this formula may be found in more complete works on field astronomy; for instance, Doolittle's "Practical Astronomy."

Z_e is the azimuth of Polaris at elongation, to be computed.
C is the correction in seconds of arc, to be subtracted from azimuth at elongation.

Table VIII gives the values of "C" of the above formula for each minute (of the interval between the instant of elongation and the instant of sighting) up to thirty minutes, and for values of Z_e from $1° 10'$ to $2° 10'$. The corrections are also given in Table V*a* near the back of the "Nautical Almanac."

This observation is a convenient one to use when the time of elongation of Polaris comes a few minutes before (or, in the morning, after) it is dark enough for the star to be seen through the telescope. The work of this observation is practically the same as that outlined in the preceding article, with the addition that the time of each pointing of the telescope at the star must be taken—at least as accurately as to the nearest minute—in order to obtain the interval "T" of Formula (31) between time of elongation and time of sighting. Though T should theoretically be a number of sidereal minutes, no appreciable error will result from using T as the number of solar minutes, as obtained directly from the computed time of culmination and the watch readings.

Outline of Observation:

Computations Preceding Field Work:

Exactly as outlined in Art. 43, page 57.

Field Work:

As near time of elongation as practicable, sight on azimuth mark with plates set at zero.

Measure the angle from the mark to the star four times by the method of repetition (the final reading on the plates should be four times the value of the angle), twice with the telescope direct and twice with it inverted. Read the plates at the first and fourth settings on the star.

Take the watch reading (standard time) to nearest half-minute at each setting on the star.

Computations Following Field Work:

Compute azimuth of star at elongation as outlined in Art. 43, page 58.

Using computed time of elongation and mean of watch readings, compute "T."

Computations Following Field Work:

Using the computed azimuth at elongation and the correction "*C*" from Table VIII, compute the azimuth of the star corresponding to the mean of the watch readings. (Subtract "*C*" from azimuth at elongation.)

Using this computed mean azimuth and the mean angle from the mark to the star, compute the azimuth of the mark.

An example of the computations and field-notes of this observation is given on pages 114 and 115.

45. Azimuth by a Circumpolar Star at any Hour Angle. This observation is one of the most precise methods for the determination of azimuth. The difficulty in its use in practice under ordinary conditions is in obtaining the standard time with sufficient accuracy. Its advantage over an observation at elongation is that the number of observations may be increased indefinitely, thereby securing greater precision.

The observation consists in measuring a series of angles between an azimuth mark and a circumpolar star, taking the time of each pointing of the telescope at the star; and from the mean of the measured angles and the azimuth of the star computed for the mean of the recorded times, computing the azimuth of the mark. In work done with great precision with large instruments several corrections need to be introduced which are too small to be considered in work done with an engineer's transit. Any of the close circumpolar stars may be used for the observation. Polaris is the best.

The mean of the recorded times of setting, read in standard time as accurately as possible,—correct within a very few seconds —may be changed to sidereal time by the methods of Arts. 26 and 30, pages 31 and 33, and Equation (24), from Art. 21, page 23, applied:

$$\text{Sid. T} = \text{R A} + \text{t} \quad . \quad . \quad . \quad . \quad . \quad (24)$$

The right ascension of the star having been obtained from the "Nautical Almanac," this formula may be used to compute the hour angle of the star at the mean of the recorded times of setting.

The latitude being known, the declination of the star having been obtained from the "Nautical Almanac," and the hour angle having been computed, the astronomical triangle may be solved for the azimuth of the star by the following formula (from Appendix A):

$$\tan Z = \frac{\sin t}{\tan \delta \,.\, \cos \phi - \sin \phi \,.\, \cos t} \quad . \quad . \quad (10)$$

If the latitude of the station is not known the altitude of the star may be read at the beginning and at the end of each set of readings of horizontal angles—one set being taken with telescope direct and one with telescope inverted—and the mean of these four altitudes, corrected for refraction, may be used in solving the astronomical triangle for azimuth by the following formula (from Appendix A):

$$\sin Z = \frac{\sin t \,.\, \cos \delta}{\cos h} \quad . \quad . \quad . \quad . \quad (11)$$

Outline of Observation:

Computations Preceding Field Work:

None.

Field Work:

By method of repetition read two sets of three or four angles each from azimuth mark to star, one set with telescope direct, and one with telescope inverted; reading the watch (*Std. T*) at the instant of each pointing at the star. Only three readings from the transit plates need be made: the value of the first angle and the reading at the end of each set.

If it is desired to solve the astronomical triangle by the second method suggested above (latitude unknown), the altitude of the star should be read at the beginning and end of each set.

Computations Following Field Work:

Compute hour angle of star at mean of observed times.

$$t = \text{Sid. T} - \text{R A} \quad . \quad . \quad . \quad . \quad . \quad . \quad . \quad . \quad . \quad (24)$$

Sid. T from mean of watch readings (*Std. T*), changed to *Sid. T*.

R A from " Nautical Almanac."

Compute azimuth of star at mean of observed times.

$$\tan Z = \frac{\sin t}{\tan \delta \,.\, \cos \phi - \sin \phi \,.\, \cos t} \quad . \quad . \quad (10)$$

t from computation above.
δ from " Nautical Almanac."
ϕ from a former observation or from an accurate map.

Compute azimuth of mark, using mean azimuth of star from above computation and mean of observed horizontal angles.

An example of the computations and field-notes of this observation is given on pages 116 and 117.

46. Azimuth by an Altitude of the Sun or of a Star. This observation consists in taking a series of sights on the sun at each of which the time, the sun's altitude, and the horizontal angle between the sun and a reference mark are read. From the mean of the observed times, the mean of the altitudes and data either known or given in the " Nautical Almanac," the mean azimuth of the sun may be computed. This mean azimuth

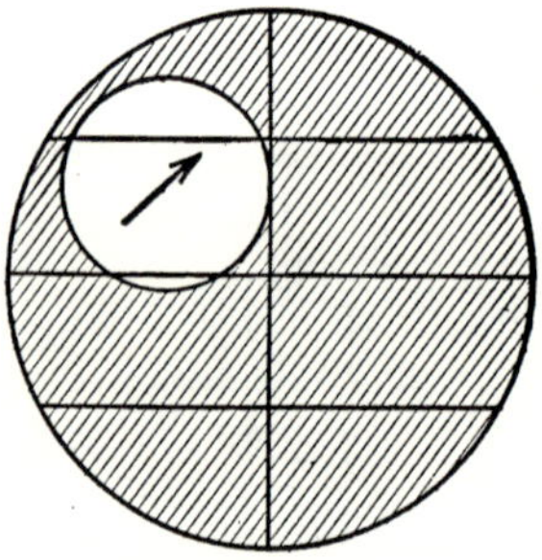

FIG. 18.

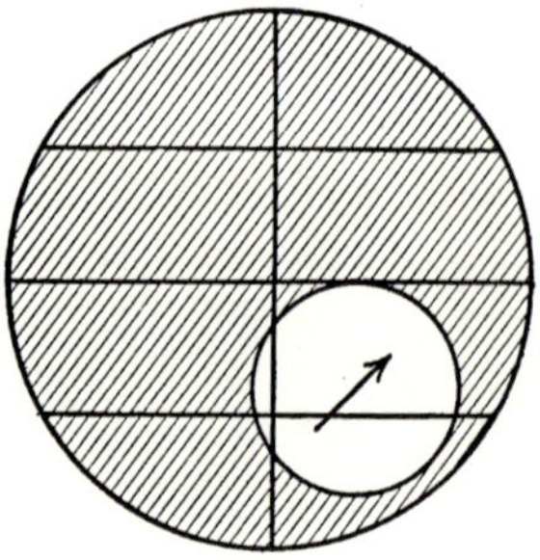

FIG. 19.

combined with the mean of the measured angles between the sun and the reference mark will give the azimuth of the mark.

The setting of the cross-hairs tangent to the sun's disc may be conveniently done in the following manner:

The transit having been centered over the station, sighted at the reference mark with the plates set at zero, the upper clamp loosened, and the telescope turned toward the sun: If the observation is being made in the forenoon the sun's disc should first be brought into the position shown in Fig. 18 and the horizontal and vertical motions clamped. The arrow shows the direction of the sun's apparent motion. The vertical cross-hair should be kept tangent to the sun's disc by use of the tangent screw of the upper horizontal motion of the transit until the upward motion of the sun has brought the disc tangent to the horizontal cross-hair. At this instant the time should be noted, and the horizontal and vertical circles read. This operation should be repeated quickly, and then the same number of settings should be made with the sun in the diagonally opposite quadrant, as shown in Fig. 19.

This time it will be convenient to keep the horizontal cross-hair tangent with the tangent screw of the vertical motion of the transit, letting the horizontal movement of the sun bring the disc tangent to the vertical cross-hair.

It is evident that the mean of the four altitudes and of the four horizontal angles will not have to be corrected for semi-diameter of the sun. If the transit has a full vertical circle the telescope should be inverted between the two sets; if not, index correction must be made. The instrument should be very carefully leveled for this observation. Care should be taken not to use one of the stadia wires for the middle cross-hair. One of the schemes for protecting the eye from the sun which are suggested in Art. 39, page 44—covering the eye-piece with either a colored glass or a prismatic eye-piece which has a colored glass or projecting the image on to a card—may be used.

It should be remembered that if a non-erecting telescope is used the direction of the sun's apparent motion will be reversed; and that if a prismatic eye-piece is used the sun's image will be turned upside down, but not left for right. However, this need not complicate matters, since it makes no difference in which quadrants the sun's image is placed if only the same number of settings is made with the sun in each of two diagonally opposite quadrants. If the observation is being made in the afternoon the relative positions of the sun's image and the cross-hairs a few seconds before becoming tangent should appear through an erecting eye-piece, as shown in Fig. 20. The procedure should be obvious.

The approximate longitude being known (scaled from a map), the sun's declination for the instant of the mean of the watch readings may be obtained by interpolation from the "Nautical Almanac." The mean of the observed altitudes, corrected for index error if necessary and for refraction and parallax, gives the altitude of the sun's center at the mean of the observed times. These, with the latitude (from a previous observation or from a map), furnish data for the solution of the astronomical triangle for azimuth by one of the formulas of Art. 12, page 13. If the latitude is not known, the hour angle of the sun's center may be obtained by changing the mean of the watch readings (standard time) to local apparent time, and used instead of the latitude in the solution of the triangle by Equation (11), (from Appendix A). Under ordinary field conditions this latter solution is likely to give less accurate results than the use of the latitude.

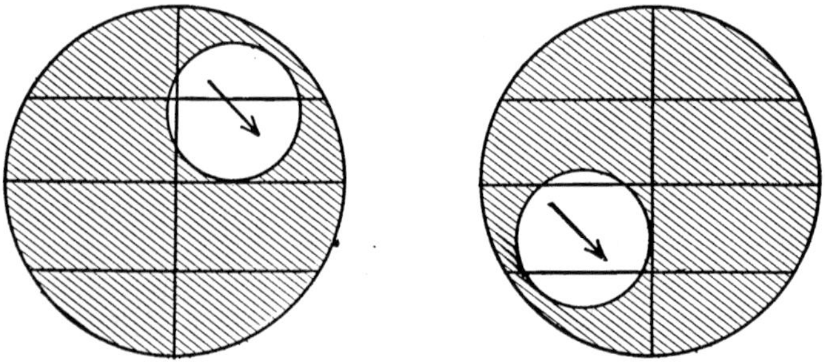

FIG. 20.

The mean azimuth of the sun thus computed, combined with the mean of the readings from the horizontal circle of the transit, will give the azimuth of the reference mark.

For good results this observation should not be made within two hours of noon or when the sun's altitude is less than 10° or 15°.

The methods of this observation may be applied equally well to an observation on a star, the star's image being bisected at each setting by both horizontal and vertical cross-hairs. The declination of a star changes so slowly that it may be considered constant for the day, so the exact time of each setting need not be taken. Of course, no parallax correction need be applied to the altitude of the star.

Outline of Observation:

Computations Preceding Field Work:

None.

Field Work:

Sight at reference mark with plates clamped at zero.

Loosen upper clamp and turn telescope toward sun.

Make two settings of cross-hairs tangent to right and lower limbs of sun's disc (if in A.M.); recording at each setting:

(1) watch reading, (2) horizontal circle reading, (3) vertical circle reading.

Repeat the settings (with telescope inverted if instrument has a full vertical circle), making the cross-hairs tangent to the left and upper limbs.

If instrument has only a vertical arc determine index correction.

Turn back to reference mark and check back-sight.

Computations Following Field Work:

Compute azimuth of sun's center at instant of mean of watch readings:

Formula from Art. 12, page 14, preferably:

$$\tan\frac{Z}{2} = \sqrt{\frac{\cos(k+\phi)\,.\,\cos(k+h)}{\sin k\,.\,\cos(k+\delta)}} \quad \ldots\ldots (8)$$

ϕ from a previous observation or from a map.

h from mean of observed altitudes, corrected for refraction and parallax, and for index error if necessary.

δ for instant of mean of watch readings by interpolation from "Nautical Almanac."

$k = \frac{1}{2}\,[270° - (\phi + h + \delta)]$.

From computed mean azimuth of sun and mean of horizontal circle readings, compute azimuth of reference mark.

It should be remembered that Z in the above formula is not always the azimuth, but is an interior angle of the astronomical triangle.

An example of the computations and field-notes of this observation is given on pages 118 and 119.

47. Azimuth by Equal Altitudes of a Star. This observation consists in marking by a stake and tack, set several hundred feet from the station, the direction of a star two or three hours before its transit (upper or lower), the altitude of the star being noted; and then marking by another stake and tack the direction of the same star at the instant when it reaches the same altitude on the opposite side of the meridian. The bisector of the angle between these two directions is the true meridian through the station.

The chief advantages of this observation are that no tables or computations and no information as to standard time are necessary; and a high degree of precision is attainable by a sufficient number of repetitions. It is inconvenient in that the two series of observations must be made at an interval of at least four to six hours.

Instead of bisecting the star with the horizontal and vertical cross-hairs and attempting to read the vertical circle during the first series of sights (before transit); it is probably better to set the vertical circle at some exact minute, so that the horizontal cross-hair is just below the star if it is moving down or just above it if it is rising, and to follow the movement of the star with the tangent screw of one of the horizontal motions (bisecting the star with the vertical wire); so that when the vertical motion of the star has brought it to the horizontal cross-hair it will be exactly at the intersection of the two. If the instrument has a vertical circle the first series of observations should be made with the telescope direct and the second series with it inverted. The refraction correction need not be considered, since its effect on the two series of observations is exactly compensating.

Any star may be used which is at a convenient altitude two or three hours before the time of its transit, and which will come to the same altitude on the other side of the meridian (four to six hours later) before daylight interferes with the observation. One of the stars of Cassiopeia or of Ursa Major may ordinarily be used.

Outline of Observation:

Computations Preceding Field Work:

None.

Field Work:

Set the vertical circle or arc to read some exact minute so that the star selected is approaching the horizontal cross-hair.

Record this reading.

Bisect the star with the vertical cross-hair and follow it with the tangent screw of one of the horizontal motions of the transit until it is at the intersection of the cross-hairs.

Line in and center a stake three or four hundred feet from the station.

Repeat this operation two or three times at intervals of ten or fifteen minutes, marking the successive stakes "A," "B," "C," etc.

When the star is approaching the same altitude on the other side of the meridian, set the vertical circle or arc to read the *last* altitude used. (If the transit has a vertical circle, have the telescope inverted.)

Bisect the star with the vertical cross-hair and follow it to the intersection of the cross-hairs as before.

Set and center a stake three or four hundred feet from the station, marking it to correspond with the *last* stake set.

Using the altitudes used in the first series of observations in the reverse order from that in which they were obtained, set as many stakes as in the first series, marking them *** "C," "B," "A."

Bisect the angle "A-station-A," and set and center a broad-topped stake three or four hundred feet from the station.

Bisect each of the other angles (B-station-B, C-station-C, etc.) and set points on the broad-topped stake beside the first.

All of these points should coincide, and a line through the station and the point of coincidence should be the meridian. If they do not coincide, the mean of their positions should be used.

Computations Following Field Work:

None.

An example of the field-notes of this observation is given on pages 120 and 121.

CHAPTER IX

OBSERVATIONS FOR TIME

48. Time by Transit of a Star. This observation consists in noting the watch time of transit of a star over the observer's meridian. The sidereal time of transit of a star may be obtained from its right ascension (taken from the "Nautical Almanac," and substituted in the equation: $Sid.\ T = R\,A + t$) and changed to standard time. The difference between this standard time and the watch reading at the instant of transit is the watch correction.

Stars to be used for the determination of time should move rapidly, and should therefore be those which are as near as practicable to the equator. Without a prismatic eye-piece—which is inconvenient for night work—stars at altitudes greater than 50° or 55° can not easily be seen with a transit; and it is not well to try to use stars whose meridian altitudes are less than 15° or 20°, if indeed the topography (surrounding mountains, etc.) does not make a higher minimum necessary.

Smaller stars than those of the fifth magnitude should not be selected as they are too difficult to observe with the ordinary transit telescope.

Before going into the field the observer should prepare a list of suitable stars, the list giving the name of the star, its magnitude, the local mean time of transit, and its approximate meridian altitude. Standard time may be used in place of local mean time if the longitude of the place is known as accurately as results are desired. If, for instance, it is desired to make the observations between eight and nine o'clock, P.M., on a certain date, these two hours changed into sidereal time (at least the approximate longitude being known) will give the limiting values of right ascension. This comes from the relation:

$$\text{Sid. T} = \text{R A} + \text{t} \quad . \quad . \quad . \quad . \quad . \quad (24)$$

t being equal to 0 hours at the time of transit.

The limiting altitudes, considered with the latitude of the observer, will determine the limiting values of declination.

Having determined the limiting values of right ascension,

declination, and magnitude, the stars may be selected from the lists given in the "Nautical Almanac," and times of transit and approximate meridian altitudes computed. Refraction may be neglected in computing the approximate altitudes.

If the line of sight is now put in the plane of the meridian, and the approximate altitude of one of the stars—the first on the list to pass the meridian—set off on the vertical arc, the star may be identified because it will cross the field of view following very nearly along the horizontal cross-hair. If the watch is keeping approximately local mean time, or if it is keeping approximately standard time and the longitude of the place is known so that approximate local mean time may be computed from the watch reading, it will assist in the identification and shorten the time of waiting.

If the transit has a vertical circle, half the observations should be made with the telescope direct and half with it inverted, the mean of the determined watch corrections being accepted. In any case, great care must be used in leveling the instrument, especially in a direction parallel to the horizontal axis. A striding level may be used to good advantage.

Outline of Observation:

Computations Preceding Field Work:

Prepare a list of four or six stars of at least fifth magnitude with meridian latitudes between 10° and 55°, and with convenient times of transit.

Meridian altitude of a southern star from:

$$h = 90° - \phi + \delta \quad . \quad . \quad . \quad . \quad . \quad . \quad . \quad . \quad . \quad (32)$$

Substitute δ with proper algebraic sign.

Local mean time of transit from:

$$\text{Sid. T} = \text{R A} + \text{t} \quad . \quad . \quad . \quad . \quad . \quad . \quad . \quad . \quad . \quad (24)$$

RA from "Nautical Almanac."

$t = 0$ hours.

Change *Sid. T* to *LMT*.

Field Work:

Having the meridian marked by two stakes, set the transit over one and sight on the other. The line of sight should now revolve (about the horizontal axis) in the plane of the meridian.

Set off on the vertical arc the approximate meridian altitude of the first star on the list (arranged in order of transit).

Field Work:

Note and record the watch reading at instant of transit of star across vertical cross-hair.

Repeat the operation for the other stars on the list, if possible making half the observations with telescope direct and half with it inverted.

Computations Following Field Work:

Determine the difference between the computed local mean time of transit and the watch reading at the instant of transit for each star observed.

Take the mean of the computed differences for the watch correction to local mean time.

An example of the computations and field notes of this observation is given on pages 122 and 123.

49. Time by Transit of the **Sun.** This observation consists in noting the watch times of transit of the west and east limbs of the sun and comparing the mean of these two watch readings with the instant of local mean time corresponding to 0 hours, local apparent time—the time of transit of the sun's center. The difference is the watch correction to local mean time. As in the work of the last article, the correction to standard time instead of to local mean time may be determined if the longitude of the place is accurately known.

Under ordinary conditions, this observation is not likely to give quite as accurate results as that of the preceding article.

Outline of Observation:

Computations Preceding Field Work:

Compute local mean time of transit of sun's center.

Change 0 hours, *LAT*, for the given date, to *LMT* by method of Art. 25, page 30.

Field Work:

Set transit over one of two stakes which mark a meridian, sight on the other, and turn telescope to approximately the meridian altitude of the sun.

Note the watch reading at the instant that each limb crosses the vertical cross-hair.

Computations Following Field Work:

Compute watch correction to local mean time by comparing mean of watch readings (taken at instants of transit of west and east limbs of sun) with computed time of transit of sun's center.

An example of the computations and field notes of this observation is given on pages 124 and 125.

The watch correction might be obtained, though probably less accurately, by solving the astronomical triangle for the hour angle of the sun's center from the data obtained as described in Art. 46, page 62, and comparing this (after having changed it from local apparent to local mean time) with the mean of the watch readings.

CHAPTER X

OBSERVATIONS FOR LONGITUDE

50. Longitude by Transportation of Timepiece. Since the difference in longitude between two places is simply the difference in local mean time, the most practicable means for determining longitude with field instruments is to compare the local mean time at one of the places (determined by one of the observations described in the last chapter) with the reading of a reliable watch or chronometer which is set to the local mean time of the other place.

If the watch is keeping standard time—the local mean time at a "standard" meridian—the difference between the watch reading and local mean time as determined by observation is the difference in longitude of the standard and local meridians. The longitude of the standard meridian with respect to the meridian of Greenwich being known, that of the local meridian may be obtained.

This method involves an approximation of the longitude for use in changing the sidereal time of transit of the stars (assuming that local mean time is to be determined by star transits, as outlined in Art. 48, page 68) to local mean time; but the error produced in this computation by a relatively large error in assumed longitude is so small that it is not likely to be of appreciable amount in work of this class. If it is in any case sufficiently large to be considered, a second determination, using for the longitude in the second computation the value obtained from the first observation, will eliminate this difficulty.

In precise geodetic work, the local mean time at two stations whose difference in longitude is being determined is compared by the electric telegraph, elaborate apparatus being used.

It is possible to determine the longitude—usually with rather indifferent precision—by an observation of the time of transit of the moon; but this method has little apparent advantage over that described above of transportation of timepiece, and its discussion is left to more complete works on field astronomy.

The necessity for the determination of longitude arises very seldom in general practice of engineering, for a locality must ordinarily be very remote whose longitude can not now be obtained from some map as accurately as it can be determined with ordinary field instruments. From a map on which distances can be scaled to the nearest mile, longitude can be obtained to the nearest minute of arc, and a determination with field instruments which is accurate to the nearest half-minute of time (seven and one half minutes of arc) is exceptionally good work.

CHAPTER XI

SUMMARY OF OBSERVATIONS

The observations which have been described in the foregoing chapters are restated here in order to summarize, for convenient reference when selecting a method of observation, the data required for each, and to state briefly some of their relative advantages.

When stating the books, etc., required for use, it is assumed that the observer has at hand a set of logarithmic and trigonometric tables.

51. Observations for Latitude.

Latitude by a Circumpolar Star at Culmination. This is one of the most precise and convenient methods of determining latitude. If Polaris is used, this book furnishes all necessary data. If any other circumpolar star is used, either the "American Ephemeris and Nautical Almanac" or the "American Nautical Almanac" will be needed from which to obtain its right ascension and declination, and in this case an approximate value of the longitude will be needed.

Latitude by Meridian Altitude of a Southern Star. The precision attainable by this method is equivalent to that of the preceding observation, but it is usually more difficult to identify a southern than a circumpolar star. (See Fig. 12, page 37.) Either the "American Ephemeris and Nautical Almanac" or the "American Nautical Almanac" will be needed for the right ascension and declination of the star. The longitude should be known approximately.

Latitude by Meridian Altitude of the Sun. The precision of this observation is probably rather inferior to that of the two preceding. It is more convenient in that it may be done in the daytime. Either the "American Ephemeris and Nautical Almanac," or the "American Nautical Almanac," or one of the pocket ephemerides published by instrument-makers, will furnish all the required data not given in this book. The longitude should be known within a few minutes of arc.

52. Observations for Azimuth.

Azimuth by a Circumpolar Star at Elongation. This is one of the most satisfactory methods of determining true azimuth under field conditions. Polaris is the best star to use unless (for a short time in the Spring and again in the Fall) its elongations occur during daylight. If Polaris is used, all necessary data, with the exception of the latitude, may be obtained from this book. For any other circumpolar star the "American Ephemeris and Nautical Almanac" or the "American Nautical Almanac" will be needed. The latitude of the station should be obtained to the nearest minute either from a reliable map or from a previous observation, and in case another star than Polaris is to be used, the longitude should be known approximately.

Azimuth by Polaris Near Elongation. This is a good observation to use when Polaris can not be seen at elongation, but can be seen within a half-hour of elongation. All necessary data may be obtained from this book, with the exception of the latitude, which should be obtained to the nearest minute from a map or from an observation, as above.

Azimuth by a Circumpolar Star at Any Hour Angle. This is a very precise method if standard time, correct within a few seconds, can be obtained. It permits a greater number of measurements of the direction of the star than either of the two preceding, and may therefore be made more precise. Either the "American Ephemeris and Nautical Almanac" or the "American Nautical Almanac" will be needed to obtain the right ascension and declination of the star, and the latitude must be obtained from a map or a previous observation. An approximate value of the longitude will be needed in computing the sidereal time.

Azimuth by an Altitude of the Sun. This method has the advantage that it may be made in the daytime and during the progress of a survey. Its precision is probably inferior to that of the observations on the stars. Either the "American Ephemeris and Nautical Almanac," or the "American Nautical Almanac," or one of the pocket solar ephemerides (preferably for Greenwich mean noon) published by the instrument makers, will furnish the required declination of the sun. The latitude and longitude should be obtained, the former to the nearest minute from a map or a previous observation, the latter less accurately from a map. Any other required data will be found in this book.

Azimuth by Equal Altitudes of a Star. This observation may be made to give a very high degree of precision, and no tables or data whatever are required. It is inconvenient in that it requires two series of observations to be made at an interval of from four to six hours.

53. Observations for Time.

Time **by Transit of Star.** This is one of the simplest and best methods of determining local time. Data from the "American Nautical Almanac," or preferably from the "American Ephemeris and Nautical Almanac," are required. The latitude and longitude should be known approximately. It is necessary that the direction of the meridian should be marked on the ground.

Time **by Transit of Sun.** This method is probably inferior in accuracy to the preceding, but it is more convenient in that it may be made in the daytime. Either the "American Ephemeris and Nautical Almanac," or the "American Nautical Almanac," or one of the pocket solar ephemerides (preferably for Greenwich mean noon) published by instrument-makers, will furnish required data. The longitude must be known approximately.

54. Observations for Longitude.

Longitude by Transportation **of Timepiece.** Since the determination of longitude by this method consists essentially in the determination of local mean time, the remarks above in regard to observations for time apply here as well.

It should be apparent that if a determination of latitude, longitude, azimuth, and time were to be made at a station where all four were entirely unknown, values of some of the quantities would have to be approximated in making observations for the others and the values obtained from the observations used for a closer approximation, until by a series of observations the values of all were obtained to the required degree of refinement.

APPENDIX A

SPHERICAL TRIGONOMETRY

Derivation of Formulas Required in Field Astronomy

A portion of the surface of a sphere bounded by arcs of three great circles is called a spherical triangle.

If the vertices of the spherical triangle be joined by straight lines to the center of the sphere, a triedral angle is formed. The face angles of the triedral angle measure the sides of the spherical triangle, and the diedral angles of the triedral angle are measures of the angles of the spherical triangle. Both sides and angles of a spherical triangle are usually measured in degrees.

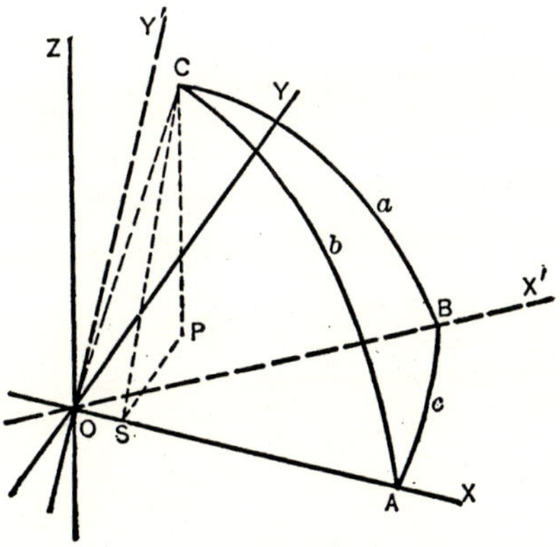

FIG. 21.

The fundamental formulas required for the solution of spherical triangles may be derived by the methods of analytic geometry, as follows:

In Fig. 21:

Let: ABC be a spherical triangle on the surface of a sphere whose center is at 0.

Then: $OA = OB = OC = r.$

Assume: O as origin of rectangular coordinates.
OX through vertex A.
OY in plane of AOB.
OZ perpendicular to plane of AOB.

Draw: CP perpendicular to plane of OX and OY.
PS perpendicular to OX.
CS.

Then: Coordinates of C are:

$$x = OS,\ y = PS,\ z = PC.$$

By construction:
Angle CSP = spherical angle A,
Angle COS = side b.

Then:

$$x = r \cdot \cos b,\ y = r \cdot \sin b \cdot \cos A,$$
$$z = r \cdot \sin b \cdot \sin A.$$

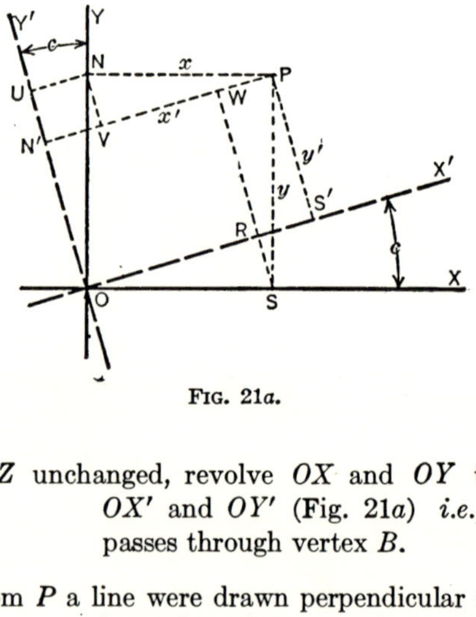

FIG. 21*a*.

Leaving OZ unchanged, revolve OX and OY to positions OX' and OY' (Fig. 21*a*) *i.e.*, until OX' passes through vertex B.

Now, if from P a line were drawn perpendicular to OX', and its point of intersection with OX' were joined to C, we should have:

$$x' = r \cdot \cos a,\ y' = -r \cdot \sin a \cdot \cos B,$$
$$z' = r \cdot \sin a \cdot \sin B.$$

The equations of transformation of coordinates in this case are:

$$x' = y \cdot \sin c \cdot + x \cdot \cos c,\quad y' = y \cdot \cos c - x \cdot \sin c,\quad z' = z.$$

Substituting in these equations the values of x, y, z, x', y', z', and dividing by r, we have:

$$\cos a = \cos b \cdot \cos c + \sin b \cdot \sin c \cdot \cos A$$
$$\sin a \cdot \cos B = \cos b \cdot \sin c - \sin b \cdot \cos c \cdot \cos A$$
$$\sin a \cdot \sin B = \sin b \cdot \sin A$$

which are the three **Fundamental Laws of Spherical Trigonometry.**

Since we are at present concerned with solutions of the astronomical triangle, lettered as shown in Fig. 22 and in other figures which show the celestial sphere, it will be convenient to have the notation of the formulas changed to correspond, as follows:

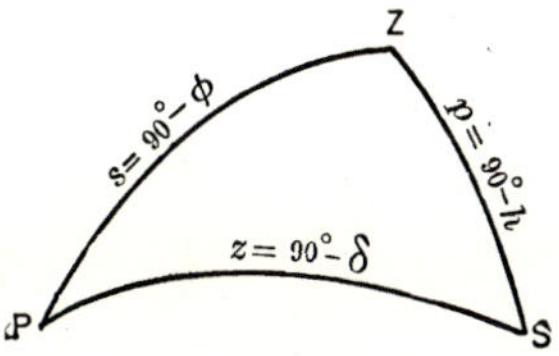

FIG. 22.

$$\cos s = \cos p \cdot \cos z + \sin p \cdot \sin z \cdot \cos S \quad . \quad . \quad . \quad (1)$$
$$\sin s \cdot \cos P = \cos p \cdot \sin z - \sin p \cdot \cos z \cdot \cos S. \quad (2)$$
$$\sin s \cdot \sin P = \sin S \cdot \sin p \quad . \quad . \quad . \quad . \quad . \quad . \quad . \quad . \quad (3)$$

From these fundamental formulas others may be derived which are more convenient for special solutions of the astronomical triangle.

To express the sines, cosines, and tangents of the half-angles of the astronomical triangle in terms of functions of the sides: From (1):

$$\cos z = \cos s \cdot \cos p + \sin s \cdot \sin p \cdot \cos Z$$

Whence:

$$\cos Z = \frac{\cos z - \cos s \cdot \cos p}{\sin s \cdot \sin p} \quad . \quad . \quad . \quad . \quad (a)$$

Subtracting both sides from 1:

$$1 - \cos Z = 1 - \frac{\cos z - \cos s \cdot \cos p}{\sin s \cdot \sin p}$$

$$= \frac{\sin s \cdot \sin p + \cos s \cdot \cos p - \cos z}{\sin s \cdot \sin p}$$

From plane trigonometry:

$$\cos (A - B) = \cos A \cdot \cos B + \sin A \cdot \sin B$$

and: $$2 \sin^2 A/2 = 1 - \cos A$$

Substituting:

$$2 \sin^2 \frac{Z}{2} = \frac{\cos (s - p) - \cos z}{\sin s \cdot \sin p}$$

From plane trigonometry:

$$\cos B - \cos A = 2 \sin \frac{A + B}{2} \cdot \sin \frac{A - B}{2} \quad . \quad . \quad \text{(b)}$$

Whence:

$$2 \sin^2 \frac{Z}{2} = \frac{2 \sin \tfrac{1}{2} [z + (s - p)] \cdot \sin \tfrac{1}{2} [z - (s - p)]}{\sin s \cdot \sin p}$$

$$\sin^2 \frac{Z}{2} = \frac{\sin \tfrac{1}{2} (z + s - p) \cdot \sin \tfrac{1}{2} (z - s + p)}{\sin s \cdot \sin p}$$

Let: $\mathbf{2k} = \mathbf{s} + \mathbf{p} + \mathbf{z} = \mathbf{270°} - (\boldsymbol{\phi} + \mathbf{h} + \boldsymbol{\delta})$

Then: $z + s - p = (s + p + z) - 2p = 2k - 2p$

$z - s + p = (s + p + z) - 2s = 2k - 2s$

Substituting:

$$\sin^2 \frac{Z}{2} = \frac{\sin (k - s) \cdot \sin (k - p)}{\sin s \cdot \sin p}$$

$$\sin \frac{Z}{2} = \sqrt{\frac{\sin (k - s) \cdot \sin (k - p)}{\sin s \cdot \sin p}} \quad \text{(c)}$$

Substituting $\phi = 90° - s$, $h = 90° - p$, $\delta = 90° - z$, in Equation (c):

$$\sin \frac{\mathbf{Z}}{\mathbf{2}} = \sqrt{\frac{\cos (\mathbf{k} + \boldsymbol{\phi}) \cdot \cos (\mathbf{k} + \mathbf{h})}{\cos \boldsymbol{\phi} \cdot \cos \mathbf{h}}} \quad . \quad . \quad (4)$$

In like manner:

$$\sin \frac{P}{2} = \sqrt{\frac{\sin (k - s) \cdot \sin (k - z)}{\sin s \cdot \sin z}}$$

$$\sin \frac{\mathbf{P}}{\mathbf{2}} = \sqrt{\frac{\cos (\mathbf{k} + \boldsymbol{\phi}) \cdot \cos (\mathbf{k} + \boldsymbol{\delta})}{\cos \boldsymbol{\phi} \cdot \cos \boldsymbol{\delta}}} \quad . \quad . \quad (5)$$

Adding both members of Equation (a) to 1:

$$1 + \cos Z = 1 + \frac{\cos z - \cos s \cdot \cos p}{\sin s \cdot \sin p}$$

$$= \frac{\cos z - (\cos s \cdot \cos p - \sin s \cdot \sin p)}{\sin s \cdot \sin p}$$

From plane trigonometry:

$$\cos (A + B) = \cos A \cdot \cos B - \sin A \cdot \sin B$$

and

$$2 \cos^2 \frac{A}{2} = 1 + \cos A$$

Substituting:

$$2 \cos^2 \frac{Z}{2} = \frac{\cos z - \cos (s + p)}{\sin s \cdot \sin p}$$

Whence, by equation (b):

$$2 \cos^2 \frac{Z}{2} = \frac{2 \sin \frac{1}{2} (s + p + z) \cdot \sin \frac{1}{2} (s + p - z)}{\sin s \cdot \sin p}$$

Again putting $2k = s + p + z$, whence: $s + p - z = 2k - 2z$:

$$\cos^2 \frac{Z}{2} = \frac{\sin k \cdot \sin (k - z)}{\sin s \cdot \sin p}$$

$$\cos \frac{Z}{2} = \sqrt{\frac{\sin k \cdot \sin (k - z)}{\sin s \cdot \sin p}}$$

or

$$\cos \frac{\mathbf{Z}}{\mathbf{2}} = \sqrt{\frac{\mathbf{\sin k \cdot \cos (k + \delta)}}{\mathbf{\cos \phi \cdot \cos h}}} \quad . \quad . \quad . \quad . \quad . \quad (6)$$

In like manner:

$$\cos \frac{P}{2} = \sqrt{\frac{\sin k \cdot \sin (k - p)}{\sin s \cdot \sin z}}$$

or

$$\cos \frac{\mathbf{P}}{\mathbf{2}} = \sqrt{\frac{\mathbf{\sin k \cdot \cos (k + h)}}{\mathbf{\cos \phi \cdot \cos \delta}}} \quad . \quad . \quad . \quad . \quad . \quad (7)$$

Dividing (4) by (6):

$$\tan \frac{Z}{2} = \sqrt{\frac{\cos (k + \phi) \cdot \cos (k + h)}{\cos \phi \cdot \cos h}} \cdot \sqrt{\frac{\cos \phi \cdot \cos h}{\sin k \cdot \cos (k + \delta)}}$$

$$\tan \frac{\mathbf{Z}}{\mathbf{2}} = \sqrt{\frac{\mathbf{\cos (k + \phi) \cdot \cos (k + h)}}{\mathbf{\sin k \cdot \cos (k + \delta)}}} \qquad (8)$$

In like manner:

$$\tan \frac{\mathbf{P}}{\mathbf{2}} = \sqrt{\frac{\mathbf{\cos (k + \phi) \cdot \cos (k + \delta)}}{\mathbf{\sin k \cdot \cos (k + h)}}} \qquad (9)$$

A convenient formula for computing the angle Z when the known data are t, ϕ, and δ, may be derived as follows:
From (3):

$$\sin p \cdot \sin Z = \sin P \cdot \sin z \quad . \quad . \quad . \quad . \qquad \textbf{(d)}$$

From (2):

$$\sin p \cdot \cos Z = \cos z \cdot \sin s - \sin z \cdot \cos s \cdot \cos P \quad \textbf{(e)}$$

Dividing (d) by (e):

$$\tan Z = \frac{\sin P \cdot \sin z}{\cos z \cdot \sin s - \sin z \cdot \cos s \cdot \cos P}$$

$$= \frac{\sin P}{\cos z \cdot \sin s - \cos s \cdot \cos P}$$

or

$$\mathbf{\tan Z} = \frac{\mathbf{\sin t}}{\mathbf{\tan \delta \cdot \cos \phi - \sin \phi \cdot \cos t}} \quad . \quad . \qquad (10)$$

A formula for computing Z when δ, t, and h are known, may be derived directly from (3):

$$\sin p \cdot \sin Z = \sin P \,.\, \sin z$$

Whence:

$$\sin Z = \frac{\sin P \cdot \sin z}{\sin p}$$

or

$$\mathbf{\sin Z} = \frac{\mathbf{\sin t \cdot \cos \delta}}{\mathbf{\cos h}} \quad . \quad . \quad . \quad . \quad . \quad . \qquad (11)$$

The formula for Z when the astronomical triangle is right-angled at S, comes directly from (3):

$$\sin s \cdot \sin Z = \sin S \cdot \sin z$$

Since $S = 90°$, $\sin S = 1$, and:

$$\sin Z = \frac{\sin z}{\sin s}$$

or

$$\mathbf{\sin Z} = \frac{\mathbf{\cos \delta}}{\mathbf{\cos \phi}} \quad . \quad . \quad . \quad . \quad . \quad . \quad . \quad . \qquad (12)$$

The formula for P under similar conditions ($S = 90°$) may be derived as follows:
From (2):

$$\sin s \cdot \cos P = \cos p \cdot \sin z - \sin p \cdot \cos z \cdot \cos S$$

Since $S = 90°$, $\cos S = 0$, and:

$$\sin s \cdot \cos P = \cos p \cdot \sin z$$

Whence:

$$\cos P = \frac{\cos p \cdot \sin z}{\sin s} \quad . \quad . \quad . \quad . \quad . \quad . \qquad \text{(f)}$$

From (1):

$$\cos s = \cos p \cdot \cos z + \sin p \cdot \sin z \cdot \cos S$$

Since $S = 90°$, $\cos S = 0$, and:

$$\cos s = \cos p \cdot \cos z$$

Multiplying the numerator of (f) by ($\cos s$), and the denominator by its equal ($\cos p \cdot \cos z$):

$$\cos P = \frac{\cos p \cdot \sin z}{\sin s} \cdot \frac{\cos s}{\cos p \cdot \cos z}$$

$$= \frac{\tan z}{\tan s}$$

or

$$\cos \mathrm{P} = \frac{\tan \phi}{\tan \delta} \quad . \quad . \quad . \quad . \quad . \quad . \quad . \qquad (13)$$

APPENDIX B

SOLAR ATTACHMENTS FOR TRANSITS

The Solar Attachment is a device which is mounted upon or beside the telescope of an engineer's transit, and is used chiefly for direct determination of the true meridian by an observation on the sun. By its use the astronomical triangle is solved mechanically, and at the end of an observation the line of sight through the transit telescope should lie in the plane of the meridian.

There are several different forms of the solar attachment made, but they are all alike in principle and differ only slightly in method of use. The essential features of all are: a **polar axis** which is perpendicular to the plane defined by the horizontal axis of the transit and the line of sight of the transit telescope; a small telescope, called the **solar telescope**, which is so mounted as to revolve about the polar axis, and which may also be set at any desired inclination to the plane of the horizontal axis and line of sight of the main telescope; and a **declination arc** or other means of measuring this inclination.

In one well-known form—the Burt Solar Attachment, shown in Fig. 23, page 85—the solar telescope is replaced by a small lens and a silver screen on which the sun's image may be thrown, thus defining a line of sight to the sun. On this form of the instrument there is a declination arc on which may be set off any desired inclination of the line of sight to the sun with respect to the plane defined by the horizontal axis and telescope of the transit.

Another common form of the attachment is that shown in Fig. 24, page 86. It has a telescope for determining the line of sight to the sun; but the inclination of this line of sight to the plane of the transit telescope and horizontal axis is determined by means of the small level tube which is mounted on the solar telescope, used in connection with the vertical arc or circle of the transit. This inclination may be set off in the following manner:

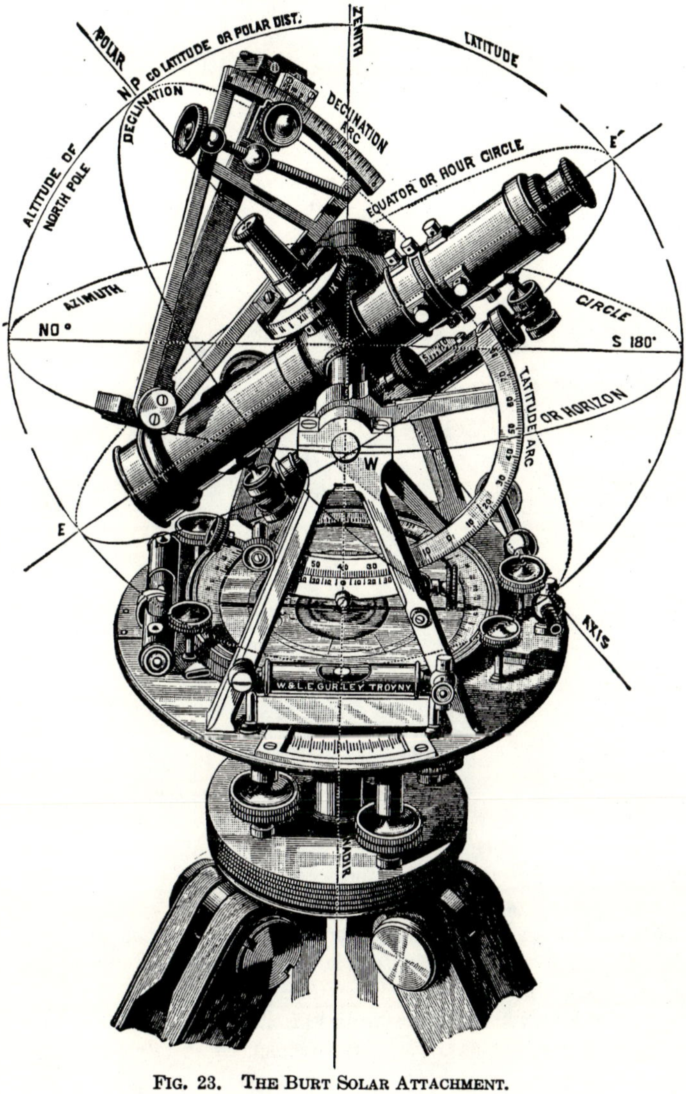

FIG. 23. THE BURT SOLAR ATTACHMENT.

First bring the solar and transit telescopes into the same vertical plane by sighting both at a distant point.

Set off the desired inclination on the vertical circle, applying index correction if necessary, and make the solar telescope horizontal by means of its own level.

The angle between the two telescopes is now equal to that set off on the vertical circle.

The objective end of the telescope should be depressed if an inclination of the line of sight of the solar telescope above the

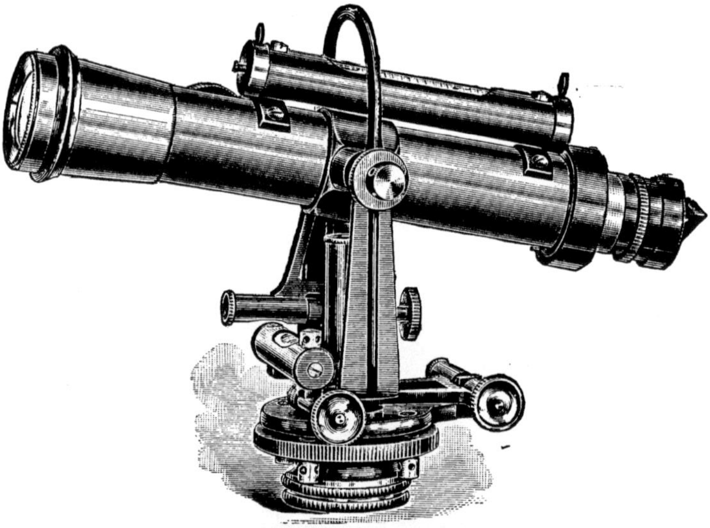

FIG. 24. THE SAEGMULLER SOLAR ATTACHMENT.

plane defined by the transit telescope and horizontal axis is desired (as for north declinations), and should be elevated for inclinations below that plane (as for south declinations).

To understand the principle on which the use of the solar attachment is based, refer to Fig. 25, page 87. Suppose the transit telescope to be so inclined that the plane defined by the line of sight and the horizontal axis of the transit coincides with the plane of the equator, and the transit to be turned so that the line of sight lies in the plane of the meridian. The polar axis will now coincide with the axis of rotation of the earth, produced; *i.e.*, it will point toward the north celestial pole. Now, assuming for the moment that the sun's declination remains constant for a day, and that the solar telescope is inclined to the plane of the equator by an amount equal to this declination, the sun's motion may be followed by the solar telescope by simply revolving the latter about the polar axis, without disturbing the rest of the instrument. Since the declination of the sun does change, its motion cannot be followed throughout the day without changing

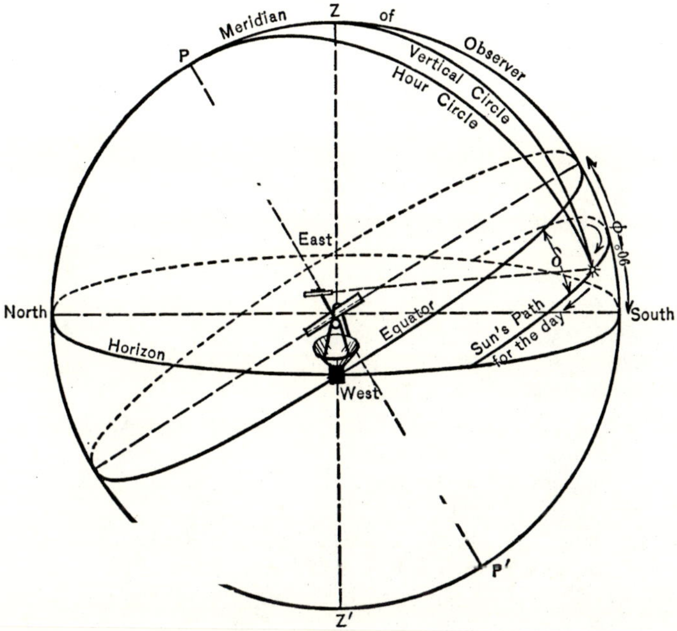

Fig. 25.

the inclination of the solar telescope; but it can be practically followed for twenty or thirty minutes, even at the seasons of most rapid change in declination—in June and December.

This cannot be done unless the polar axis coincides with the axis of the celestial sphere (points, in the northern hemisphere, to the north celestial pole) and the telescope lies in the plane of the meridian. This condition furnishes the principle on which the work of an observation is based.

The declination of the sun, computed for the time at which the observation is to be made, must be corrected for refraction. The accuracy with which the declination can be set off does not warrant the use of a parallax correction. To obtain the "apparent" from the computed or true declination, the refraction correction is always *added algebraically,* regarding north declinations as plus and south declinations as minus. The amount of this correction may be obtained from Table III, page 91, using as the altitude the approximate measured altitude of the sun; or it may be obtained more easily from tables such as are published by the different instrument-makers for use with solar attachments; in which the refraction corrections are given for various latitudes, declinations, and hour angles of the sun (before or after transit).

The observation for azimuth with the solar attachment may now be outlined as follows:

Set up the transit over a centered stake.

Incline the solar telescope to the plane of the transit telescope and horizontal axis by an amount equal to the declination of the sun—" corrected " for refraction—computed for the day and hour of the observation.

Make sure that the transit is accurately leveled.

Without disturbing the relative inclination of the two telescopes incline the transit telescope by an amount equal to the co-latitude of the place (making the polar axis point toward the north celestial pole); applying index correction, if necessary, to obtain the proper setting for the vertical arc or circle.

By revolving the transit about its vertical axis and the solar telescope about the polar axis, bring the sun's image into the center of the square formed by the four cross-hairs of the solar telescope or by the four lines on the silver plate of the attachment which has no telescope. Finish the setting with the tangent screws. A magnifying glass should be used in checking the position of the image on the silver plate.

Only one position can be found in which this setting can be properly made. When this position is found it should be possible to follow the motion of the sun for several minutes, keeping the image in the little square, by simply turning the solar telescope about the polar axis, and without otherwise disturbing the instrument.

The line of sight of the transit telescope should now be in the plane of the meridian; and a stake may be lined in and centered, which, with the stake under the instrument, will define the direction of the meridian.

Also, the approximate local apparent time should now be

indicated on a graduated circle which is near the base of the polar axis on some types of the instrument, and is perpendicular to the polar axis.

Most of the companies making solar attachments publish yearly a little pamphlet containing, in addition to a solar ephemeris and other useful tables, detailed directions for the adjustment and use of their own instruments. Some companies publish these directions in a pamphlet or book separate from their solar ephemeris. The reader is referred to these publications for more particular information in regard to the several solar attachments now on the market.

Observations for latitude and for time may be made with the solar attachment, but it is believed that its use in these observations presents but little advantage over the methods given in Chapters VII and IX.

The chief advantage of the solar attachment in observations for azimuth is that the direction of the meridian may be obtained directly, at any time of day when the sun is visible and with a minimum of computation. For good results, however, observations should not be made when the altitude of the sun is less than 10° or 15° or within an hour of noon.

The solar attachment may also be used in a series of observations for running lines by true bearings much as a compass is used for running lines by magnetic bearings. This work is explained in the publications of the instrument-makers and in most texts on plane surveying.

In regard to the accuracy attainable, it is claimed by some that results accurate within a quarter of a minute may be obtained in observations for azimuth with the solar attachment. It is believed, however, that the nearest minute of arc represents about as high a degree of accuracy as is likely to be consistently realized.

TABLES

TABLE I

CONVERSION OF SIDEREAL INTO MEAN SOLAR TIME

Corrections to be Subtracted from a Sidereal Time Interval

Sid. Hrs.	Corr. m s	Sid. Min.	Corr. s	Sid. Min.	Corr. s	Sid. Sec.	Corr. s	Sid. Sec.	Corr. s
1.......	0 09.83	1	0.16	31	5.08	1	0.003	31	0.08
2.......	0 19.66	2	0.33	32	5.24	2	0.005	32	0.09
3.......	0 29.49	3	0.49	33	5.41	3	0.008	33	0.09
4.......	0 39.32	4	0.66	34	5.57	4	0.011	34	0.09
5.......	0 49.15	5	0.82	35	5.73	5	0.014	35	0.10
6.......	0 58.98	6	0.98	36	5.90	6	0.02	36	0.10
7.......	1 08.81	7	1.15	37	6.06	7	0.02	37	0.10
8.......	1 18.64	8	1.31	38	6.23	8	0.02	38	0.10
9.......	1 28.47	9	1.47	39	6.39	9	0.03	39	0.11
10.......	1 38.30	10	1.64	40	6.55	10	0.03	40	0.11
11.......	1 48.13	11	1.80	41	6.72	11	0.03	41	0.11
12.......	1 57.96	12	1.97	42	6.88	12	0.03	42	0.12
13.......	2 07.78	13	2.13	43	7.05	13	0.04	43	0.12
14.......	2 17.61	14	2.29	44	7.21	14	0.04	44	0.12
15.......	2 27.44	15	2.46	45	7.37	15	0.04	45	0.12
16.......	2 37.27	16	2.62	46	7.54	16	0.04	46	0.13
17.......	2 47.10	17	2.79	47	7.70	17	0.05	47	0.13
18.......	2 56.93	18	2.95	48	7.86	18	0.05	48	0.13
19.......	3 06.76	19	3.11	49	8.03	19	0.05	49	0.13
20.......	3 16.59	20	3.28	50	8.19	20	0.06	50	0.14
21.......	3 26.42	21	3.44	51	8.36	21	0.06	51	0.14
22.......	3 36.25	22	3.60	52	8.52	22	0.06	52	0.14
23.......	3 46.08	23	3.77	53	8.68	23	0.06	53	0.15
24.......	3 55.91	24	3.93	54	8.85	24	0.07	54	0.15
		25	4.10	55	9.01	25	0.07	55	0.15
		26	4.26	56	9.17	26	0.07	56	0.15
		27	4.42	57	9.34	27	0.07	57	0.16
		28	4.59	58	9.50	28	0.08	58	0.16
		29	4.75	59	9.67	29	0.08	59	0.16
		30	4.92	60	9.83	30	0.08	60	0.16

TABLE II

CONVERSION OF MEAN SOLAR INTO SIDEREAL TIME

Corrections to be Added to a Mean Solar Time Interval

Mean Solar Hrs.	Corr. m s	Mean Solar Min.	Corr. s	Mean Solar Min.	Corr. s	Mean Solar Sec.	Corr. s	Mean Solar Sec.	Corr. s
1.......	0 09.86	1	0.16	31	5.09	1	0.003	31	0.09
2.......	0 19.71	2	0.33	32	5.26	2	0.005	32	0.09
3.......	0 29.57	3	0.49	33	5.42	3	0.008	33	0.09
4.......	0 39.43	4	0.66	34	5.59	4	0.011	34	0.09
5.......	0 49.28	5	0.82	35	5.75	5	0.014	35	0.10
6.......	0 59.14	6	0.99	36	5.91	6	0.02	36	0.10
7.......	1 09.00	7	1.15	37	6.08	7	0.02	37	0.10
8.......	1 18.85	8	1.31	38	6.24	8	0.02	38	0.10
9.......	1 28.71	9	1.48	39	6.41	9	0.03	39	0.11
10.......	1 38.57	10	1.64	40	6.57	10	0.03	40	0.11
11.......	1 48.42	11	1.81	41	6.74	11	0.03	41	0.11
12.......	1 58.28	12	1.97	42	6.90	12	0.03	42	0.12
13.......	2 08.13	13	2.14	43	7.06	13	0.04	43	0.12
14.......	2 17.99	14	2.30	44	7.23	14	0.04	44	0.12
15.......	2 27.85	15	2.46	45	7.39	15	0.04	45	0.12
16.......	2 37.70	16	2.63	46	7.56	16	0.04	46	0.13
17.......	2 47.56	17	2.79	47	7.72	17	0.05	47	0.13
18.......	2 57.42	18	2.96	48	7.89	18	0.05	48	0.13
19.......	3 07.27	19	3.12	49	8.05	19	0.05	49	0.13
20.......	3 17.13	20	3.29	50	8.21	20	0.06	50	0.14
21.......	3 26.99	21	3.45	51	8.38	21	0.06	51	0.14
22.......	3 36.84	22	3.61	52	8.54	22	0.06	52	0.14
23.......	3 46.70	23	3.78	53	8.71	23	0.06	53	0.15
24.......	3 56.56	24	3.94	54	8.87	24	0.07	54	0.15
		25	4.11	55	9.04	25	0.07	55	0.15
		26	4.27	56	9.20	26	0.07	56	0.15
		27	4.44	57	9.36	27	0.07	57	0.16
		28	4.60	58	9.53	28	0.08	58	0.16
		29	4.76	59	9.69	29	0.08	59	0.16
		30	4.93	60	9.86	30	0.08	60	0.16

TABLE III

MEAN REFRACTION

Corrections to be Subtracted from Apparent (Observed) Altitudes

Barometer: 29.6 inches Temperature: 49° F.

Apparent Altitude	Mean Refraction	Apparent Altitude	Mean Refraction
10°	5′ 16″	38°	1′ 14″
11	4 49	39	1 11
12	4 25	40	1 09
13	4 05	41	1 06
14	3 47	42	1 04
15	3 32	43	1 02
16	3 19	44	1 00
17	3 07	45	0 58
18	2 56	46	0 56
19	2 46	47	0 54
20	2 37	48	0 52
21	2 29	49	0 50
22	2 22	50	0 48
23	2 15	51	0 47
24	2 09	52	0 45
25	2 03	53	0 44
26	1 58	54	0 42
27	1 53	55	0 40
28	1 48	56	0 39
29	1 44	57	0 38
30	1 40	58	0 36
31	1 36	59	0 35
32	1 32	60	0 33
33	1 29	65	0 27
34	1 25	70	0 21
35	1 22	75	0 16
36	1 19	80	0 10
37	1 17	85	0 05
		90	0 00

For temperatures other than 49° F., the mean refraction may be multiplied by the following factors:

For 20°: 1.060, for 40°: 1.017, for 60°: 0.978, for 80°: 0.942, for 88°: 0.929.

These mean refractions are based on Bessel's "Refraction Tables."

TABLE IV

SUN'S PARALLAX AND SEMI-DIAMETER

SUN'S PARALLAX

Altitude	Parallax	Altitude	Parallax
0°	9″	50°	6″
10	9	60	4
20	8	70	3
30	8	80	2
40	7	90	0

SUN'S SEMI-DIAMETER

Date	Semi-Diameter	Date	Semi-Diameter
January 1.......	16′ 18″	July 1..........	15′ 46″
February 1......	16 16	August 1........	15 47
March 1........	16 10	September 1.....	15 53
April 1..........	16 02	October 1.......	16 01
May 1..........	15 54	November 1.....	16 09
June 1..........	15 48	December 1.....	16 15

TABLE V

LOCAL MEAN (ASTRONOMICAL) TIME OF THE CULMINATIONS AND ELONGATIONS OF POLARIS IN THE YEAR 1915

With Corrections for Referring the Tabular Quantities to Other Years

(Computed for latitude 40° north and longitude 90° or 6 h west of Greenwich)

Date 1915	East Elongation h	m	Upper Culmination h	m	West Elongation h	m	Lower Culmination h	m
January 1	0	51.7	6	46.9	12	42.1	18	44.9
January 15	23	52.5	5	51.6	11	46.8	17	49.6
February 1	22	45.3	4	44.5	10	39.7	16	42.5
February 15	21	50.1	3	49.2	9	44.4	15	47.2
March 1	20	54.8	2	54.0	8	49.2	14	52.0
March 15	19	59.6	1	58.8	7	54.0	13	56.8
April 1	18	52.7	0	51.9	6	47.1	12	49.9
April 15	17	57.7	23	52.9	5	52.0	11	54.8
May 1	16	54.8	22	50.0	4	49.2	10	52.0
May 15	15	59.9	21	55.1	3	54.2	9	57.0
June 1	14	53.3	20	48.5	2	47.6	8	50.4
June 15	13	58.5	19	53.7	1	52.8	7	55.6
July 1	12	55.9	18	51.1	0	50.2	6	53.0
July 15	12	01.1	17	56.3	23	51.5	5	58.2
August 1	10	54.5	16	49.7	22	44.9	4	51.7
August 15	9	59.8	15	55.0	21	50.2	3	56.9
September 1	8	53.2	14	48.4	20	43.6	2	50.3
September 15	7	58.3	13	53.5	19	48.7	1	55.4
October 1	6	55.5	12	50.7	18	45.9	0	52.7
October 15	6	00.6	11	55.8	17	51.0	23	53.8
November 1	4	53.7	10	48.9	16	44.1	22	46.9
November 15	3	58.6	9	53.8	15	49.0	21	51.8
December 1	2	55.6	8	50.8	14	46.0	20	48.8
December 15	2	00.4	7	55.6	13	50.8	19	53.6

Corrections on pages 98 and 99

A. To refer the above tabular quantities to years other than 1915:

		m
For year	1916, add	1.6 up to March 1
	1916, subtract	2.3 on and after March 1
	1917, subtract	0.7
	1918, add	0.9
	1919, add	2.5
	1920, add	4.0 up to March 1
	1920, add	0.1 on and after March 1
	1921, add	1.6
	1922, add	3.1
	1923, add	4.5
	1924, add	5.9 up to March 1
	1924, add	2.0 on and after March 1
	1925, add	3.3
	1926, add	4.6
	1927, add	5.9
	1928, add	7.2 up to March 1.
	1928, add	3.3 on and after March 1.

B. To refer to any calendar day other than the first and fifteenth of each month, *subtract* the quantities below from the tabular quantity for the preceding date.

Day of Month	Minutes	No. of Days Elapsed
2 or 16	3.9	1
3 17	7.8	2
4 18	11.8	3
5 19	15.7	4
6 20	19.6	5
7 21	23.5	6
8 22	27.4	7
9 23	31.4	8
10 24	35.3	9
11 25	39.2	10
12 26	43.1	11
13 27	47.0	12
14 28	51.0	13
29	54.9	14
30	58.8	15
31	62.7	16

Continued on page 99

C. To refer the table to Standard time and to the civil or common method of reckoning:

(a) *Add* to the tabular quantities four minutes for every degree of longitude the place is west of the standard meridian and *subtract* when the place is east of the standard meridian.

(b) The astronomical day begins twelve hours after the civil day, *i.e.*, begins at noon on the civil day of the same date, and is reckoned from zero to twenty-four hours. Consequently, an astronomical time less than twelve hours refers to the same civil day, whereas an astronomical time greater than twelve hours refers to the morning of the next civil day.

It will be noticed that for the tabular year two eastern elongations occur on January 14 and two western elongations on July 13. There are also two upper culminations on April 14 and two lower culminations on October 14. The lower culmination either follows or precedes the upper culminating by 11^h $58^m.0$.

D. To refer to any other than the tabular latitude between the limits of 10° and 50° north:

Add to the time of west elongation $0^m.10$ for every degree south of 40° and *subtract* from the time of west elongation $0^m.16$ for every degree north of 40°. Reverse these operations for correcting time of east elongation.

E. To refer to any other than the tabular longitude:

Add $0^m.16$ for each 15° east of the ninetieth meridian and *subtract* $0^m.16$ for each 15° west of the ninetieth meridian.

Table V and the accompanying supplementary tables and rules have been kindly furnished for this book by the Superintendent of the United States Coast and Geodetic Survey.

TABLE VI

MEAN DECLINATIONS OF POLARIS

For January 1 of the Years from 1915 to 1928

Year	Mean Declination	Year	Mean Declination
1915	+88° 51.11′	1922	+88° 53.27′
1916	88 51.42	1923	88 53.57
1917	88 51.73	1924	88 53.88
1918	88 52.03	1925	88 54.19
1919	88 52.34	1926	88 54.49
1920	88 52.65	1927	88 54.80
1921	88 52.96	1928	88 55.10

The above table is based on data obtained from the "American Ephemeris and Nautical Almanac." The apparent declination of Polaris for any day in the year may be taken from the "American Ephemeris and Nautical Almanac" or from the "American Nautical Almanac." For 1916 the apparent declination decreases from 88° 51′ 50″.80 on January 1 to 88° 51′ 21″.57 on June 27, and then increases to 88° 52′ 10″.50 on December 31.

When a possible error of about half a minute in declination is too large to be allowed the apparent declination should be obtained from the Almanac for the given date.

TABLE VII

AZIMUTH OF POLARIS WHEN AT ELONGATION

For Any Year Between 1915 and 1928

Latitude	1915	1916	1917	1918	1919	1920	1921
10°	1° 10.0′	1° 09.6′	1° 09.3′	1° 09.0′	1° 08.7′	1° 08.4′	1° 08.1′
11	10.2	09.9	09.6	09.2	08.9	08.6	08.3
12	10.4	10.1	09.8	09.5	09.2	08.9	08.6
13	10.7	10.4	10.1	09.8	09.4	09.1	08.8
14	11.0	10.7	10.4	10.0	09.7	09.4	09.1
15	11.3	11.0	10.7	10.4	10.0	09.7	09.4
16	11.7	11.4	11.0	10.7	10.4	10.1	09.8
17	12.0	11.7	11.4	11.1	10.8	10.4	10.1
18	12.4	12.1	11.8	11.5	11.1	10.7	10.5
19	12.8	12.5	12.2	11.9	11.6	11.2	10.9
20	13.3	13.0	12.7	12.3	12.0	11.7	11.4
21	13.8	13.5	13.1	12.8	12.5	12.2	11.8
22	14.3	14.0	13.6	13.3	13.0	12.6	12.3
23	14.8	14.5	14.2	13.8	13.5	13.2	12.8
24	15.4	15.1	14.7	14.4	14.1	13.7	13.4
25	16.0	15.7	15.3	15.0	14.7	14.3	14.0
26	16.6	16.3	16.3	15.6	15.3	14.9	14.7
27	17.3	17.0	16.6	16.3	15.9	15.6	15.2
28	18.0	17.7	17.3	17.0	16.6	16.3	15.9
29	18.8	18.4	18.1	17.7	17.4	17.0	16.6
30	19.6	19.2	18.8	18.5	18.1	17.8	17.4
31	20.4	20.0	19.7	19.3	18.9	18.6	18.2
32	21.2	20.9	20.5	20.1	19.8	19.4	19.1
33	22.1	21.8	21.4	21.0	20.7	20.3	19.9
34	23.1	22.7	22.4	22.0	21.6	21.2	20.9
35	24.1	23.7	23.3	23.0	22.6	22.2	21.8
36	25.2	24.8	24.4	24.0	23.6	23.3	22.9
37	26.3	25.9	25.3	25.1	24.7	24.3	24.0
38	27.4	27.0	26.6	26.2	25.9	25.5	25.1
39	28.6	28.2	27.8	27.5	27.1	26.7	26.3
40	29.9	29.5	29.1	28.7	28.3	27.9	27.5
41	31.3	30.9	30.4	30.0	29.6	29.1	28.8
42	32.7	32.3	31.9	31.5	31.0	30.6	30.2
43	34.2	33.8	33.4	32.9	32.5	32.1	31.8
44	35.8	35.3	34.9	34.5	34.1	33.6	33.2
45	37.4	37.0	36.6	36.1	35.7	35.3	34.8
46	39.2	38.7	38.3	37.8	37.4	37.0	36.5
47	41.0	40.6	40.1	39.7	39.2	38.8	38.3
48	43.0	42.5	42.0	41.6	41.1	40.7	40.2
49	45.0	44.5	44.1	43.6	43.1	42.7	42.2
50	1° 47.2′	1° 46.7′	1° 46.2′	1° 45.7	1° 45.3′	1° 44.8′	1° 44.3′

Continued on page 102　　*Corrections on page 103*

TABLE VII—*Continued*

AZIMUTH OF POLARIS WHEN AT ELONGATION

For Any Year Between 1915 and 1928

Latitude	1922	1923	1924	1925	1926	1927	**1928**
10°..........	1° 07.8′	1° 07.4′	1° 07.2′	1° 06.8′	1° 06.5′	1° 06.2′	1° 05.9′
11..........	08.0	07.7	07.4	07.0	06.7	06.4	06.1
12..........	08.2	07.9	07.6	07.3	07.0	06.7	06.4
13..........	08.5	08.2	07.8	07.6	07.2	06.9	06.6
14..........	08.8	08.5	08.2	07.8	07.5	07.2	06.9
15..........	09.1	08.8	08.5	08.1	07.8	07.5	07.2
16..........	09.4	09.1	08.8	08.5	08.2	07.8	07.5
17..........	09.8	09.5	09.2	08.8	08.5	08.2	07.9
18..........	10.2	09.8	09.5	09.2	08.9	08.6	08.2
19..........	10.6	10.2	09.9	09.6	09.3	09.0	08.6
20..........	11.0	10.7	10.4	10.0	09.7	09.4	09.1
21..........	11.5	11.2	10.8	10.5	10.2	09.8	09.5
22..........	12.0	11.6	11.3	11.0	10.6	10.3	10.0
23..........	12.5	12.2	11.8	11.5	11.2	10.8	10.5
24..........	13.0	12.7	12.4	12.0	11.7	11.4	11.0
25..........	13.6	13.3	13.0	12.6	12.3	11.9	11.6
26..........	14.2	13.9	13.6	13.2	12.9	12.5	12.2
27..........	14.9	14.6	14.2	13.9	13.5	13.2	12.8
28..........	15.6	15.2	14.9	14.6	14.2	13.8	13.5
29..........	16.3	16.0	15.6	15.2	14.9	14.6	14.2
30..........	17.0	16.7	16.4	16.0	15.6	15.3	14.9
31..........	17.9	17.5	17.2	16.8	16.4	16.1	15.7
32..........	18.7	18.3	18.0	17.6	17.2	16.9	16.5
33..........	19.6	19.2	18.8	18.5	18.1	17.8	17.4
34..........	20.5	20.1	19.8	19.4	19.0	18.6	18.3
35..........	21.5	21.1	20.7	20.4	20.0	19.6	19.2
36..........	22.5	22.1	21.7	21.4	21.0	20.6	20.2
37..........	23.6	23.2	22.8	22.4	22.0	21.6	21.3
38..........	24.7	24.3	23.9	23.5	23.2	22.8	22.4
39..........	25.8	25.5	25.1	24.7	24.3	23.9	23.5
40..........	27.1	26.7	26.3	25.9	25.5	25.1	24.7
41..........	28.4	28.0	27.6	27.2	26.8	26.4	26.0
42..........	29.8	29.4	29.0	28.6	28.2	27.8	27.3
43..........	31.2	30.8	30.4	30.0	29.6	29.1	28.7
44..........	32.8	32.4	31.9	31.5	31.1	30.6	30.2
45..........	34.4	34.0	33.5	33.1	32.6	32.2	31.8
46..........	36.1	35.6	35.2	34.8	34.3	33.9	33.4
47..........	37.9	37.4	37.0	36.5	36.1	35.6	35.2
48..........	39.8	39.3	38.8	38.4	37.9	37.4	37.0
49..........	41.7	41.3	40.8	40.3	39.9	39.4	38.9
50..........	1° 43.8′	1° 43.4′	1° 42.9′	1° 42.4′	1° 41.9′	1° 41.4′	1° 41.1′

Corrections on page 103

The preceding table was computed with the mean declination of Polaris for each year. A more accurate result will be had by applying to the tabular values the following corrections, which depend on the difference of the mean and the apparent place of the star. The deduced azimuth will, in general, be correct within 0′.3.

For Middle of	Correction	For Middle of	Correction
January	−0.5′	July	+0.2′
February	−0.4	August	+0.1
March	−0.3	September	−0.1
April	0.0	October	−0.4
May	+0.1	November	−0.6
June	+0.2	December	−0.8

Table VII and the accompanying supplementary table have been kindly furnished for this book by the Superintendent of the United States Coast and Geodetic Survey.

TABLE VIII

CORRECTIONS FOR OBTAINING AZIMUTH OF POLARIS WHEN NEAR ELONGATION FROM AZIMUTH AT ELONGATION

Interval from Elongation in Minutes	Azimuth at Elongation								
	1° 00′	1° 10′	1° 20′	1° 30′	1° 40′	1° 50′	2° 00′	2° 10′	2° 20′
0	0.0″	0.0″	0.0″	0.0″	0.0″	0.0″	0.0″	0.0″	0.0″
1	0.0	0.0	0.0	0.1	0.1	0.1	0.1	0.1	0.1
2	0.1	0.2	0.2	0.2	0.2	0.3	0.3	0.3	0.3
3	0.3	0.4	0.4	0.5	0.5	0.6	0.6	0.7	0.7
4	0.5	0.6	0.7	0.8	0.9	1.0	1.1	1.2	1.3
5	1	1	1	1	1	2	2	2	2
6	1	1	2	2	2	2	3	3	3
7	2	2	2	3	3	3	3	4	4
8	2	3	3	3	4	4	4	5	5
9	3	3	4	4	5	5	6	6	7
10	3	4	5	5	6	6	7	7	8
11	4	5	6	6	7	8	8	9	10
12	5	6	7	7	8	9	10	11	12
13	6	7	8	9	10	11	12	13	14
14	7	8	9	10	11	12	13	15	16
15	8	9	10	12	13	14	15	17	18
16	9	10	12	13	15	16	18	19	21
17	10	12	13	15	17	18	20	22	23
18	11	13	15	17	19	20	22	24	26
19	12	14	17	19	21	23	25	27	29
20	14	16	18	21	23	25	27	30	32
21	15	18	20	23	25	28	30	33	35
22	17	19	22	25	27	30	33	36	39
23	18	21	24	27	30	33	36	39	42
24	20	23	26	29	33	36	40	43	46
25	21	25	29	32	36	39	43	46	50
26	23	27	31	35	39	42	46	50	54
27	25	29	33	38	42	46	50	54	58
28	27	31	36	40	45	49	54	58	63
29	28	34	38	43	48	53	58	62	67
30	30	36	41	46	51	57	62	67	72

GREEK ALPHABET

Letters	*Name*	*Letters*	*Name*
Α, α	Alpha	Ν, ν	Nu
Β, β	Beta	Ξ, ξ	Xi
Γ, γ	Gamma	Ο, ο	Omicron
Δ, δ	Delta	Π, π	Pi
Ε, ϵ	Epsilon	Ρ, ρ	Rho
Ζ, ζ	Zeta	Σ, σ, ς	Sigma
Η, η	Eta	Τ, τ	Tau
Θ, θ	Theta	Υ, υ	Upsilon
Ι, ι	Iota	Φ, φ	Phi
Κ, κ	Kappa	Χ, χ	Chi
Λ, λ	Lambda	Ψ, ψ	Psi
Μ, μ	Mu	Ω, ω	Omega

Latitude by Polaris at Lower Culmination.
Station: Δ1, N. end of Base Line.

λ = 73°00′ W, approx.

Observed Double Altitudes:

	(1)	86°17′20″
	(2)	86 17 10
	(3)	86 17 10
		3) 40
		86-17-13
Index Corr.:	—	10
	2)	86 17 03
		43 08 32
Refr. Corr.:	—	1 02
	h:	43°07′30″

A.B.C., Observer.	"K. + E." Sextant;
D.E.F., Recorder.	Vernier: 00' 10"

May 20, 1916.

Time of Lower Culm of Polaris:

Sid. T = RA + t

$= 1^h 28^m 57^s + 12^h 00^m 00^s = 13^h 28^m 57^s$

A, at Gh. M.N., May 20: 3 51 14				
Increase for λ (4ʰ 52ᵐ): 0 48				
An,		3	52	02
Sid. Int. since M.N.:		9	36	55
Corr., Sid. to Solar:	−		01	35
L.M.T., 73°00' W., Astr., May 20		9	35	20
" " , Civil, May 20, P.M.,		9	35	20
Corr., Local to Std.:	−		08	00
Std. T (Eastern) of L.C.: P.M.,		9	27	20 ✓

From Table V:

Std. T of L.C.: May 20, P.M., $9^h 27.3^m$ ✓

Computation of Latitude:

$\phi = h + (90° - \delta)$

$= 43° 07' 30" + (90° - 88° 51' 26")$

$= 44° 16' 04"$

or 44° 16' 00" to nearest 10".

Latitude by Meridian Altitude of
α Virginis (Spica).

Station: Near Eng'ng. Bldg., Iowa City, Ia.

$\lambda = 91° 31' 30'' W.$

Observed Altitudes:

Telescope Direct:	37.°62
" Inverted:	37.63
2)	75.25
Mean:	37.625
Refr. Corr.: −	0.021
h:	37.604

A.H.H., Observer.
R.B.H. Recorder.

"K + E" Transit
Vern. of Vert. Circle: 0°.01

June 3, 1916.

Time of Transit of α Virginis (Spica):

Sid. T = RA + t

$= 13^h\ 20^m\ 49^s + 0^h\ 00^m\ 00^s = 13^h\ 20^m\ 49^s$

A for Gh M.N., June 3: 4 46 26				
Increase for λ ($6^h 06^m 06^s$): 01 00				
An		4	47	26
Sid. Int. since M.N.:		8	33	23
Corr., Sid. to Solar:	−		1	24
LMT, 91-31-30 W, Astr., June 3:		8	31	59
" " , Civil, " :	P.M.	8	31	59
Corr., Local to Std.:	+		06	06
Std. T, Central, June 3:	P.M.	8	38	05

Computation of Latitude:

$\phi = 90° - (h - \delta)$

$= 90° - [37°.604 - (-10°.729)]$

$= 41°.667$

or 41°.67, to nearest 0°.01.

Latitude by Meridian Altitude of the Sun.
Station: Δ1, N. end of Base Line.

λ =	73°00′W
Observed Merid. Alt. of ☉	66°05′
Index Corr.:	00
	66 05
Refr. Corr.: −	00 26″
	66 04 34
Semi-diam. Corr.: +	15 50
	66 20 24
Parallax Corr. +	00 03
h:	66°20′27″

A.B.C., Observer. D.E.F., Recorder.	Gurley Transit. Vern. of Vert Circle: 01' May 22, 1916.

Time of Transit of Sun's Center:

LAT, 73°00'W, May 22,		$0^h\ 00^m\ 00^s$
λ W of Gh.:	$4^h\ 52^m\ 00^s$	
Since Gh. App. N.:	4 52 00	
E:		− 03 31
LMT, 73°00'W, Astr., May 21:		23 56 29
" , 75°00'W, " , " ":		23 48 29
Std. T, Eastern, May 22: A.M.		11 48 29

"E" was taken from Am. Naut Almanac, and was assumed to be the same at Gh. App. N. as at Gh. M. N.

Computation of Latitude:

$\phi = 90° - (h - \delta)$

$= 90° - [66°20'27'' - (+20°36'36'')]$

$= 44°16'09''$

or <u>44°16'</u>, to nearest minute.

Azimuth by Polaris at East Elongation.
Station: Stk. near N E cor. of Washington and Madison Sts., Iowa City, Iowa.

$\phi = 41°40'$, $\lambda = 91°31'30''$

Horiz. ∠s betw. Azimuth Mark and Star:

No. Rep.	Horiz. Circle Readings		
	'A' Vern.	'B' Vern.	Mean
On Mk.	00°00'00"	00' 00"	00' 00"
1 D	15 11 00	11 30	
2 D	30 22 30	22 30	
2 R	60 45 30	45 00	45 15
	Arc passed over:		60°45'15"
	Mean Angle:		15 11 19

Bearing of Azimuth Mark:

15° 11' 19"
1 31 57
13 39 22

or, N13° 39' 20" W, to nearest 10".

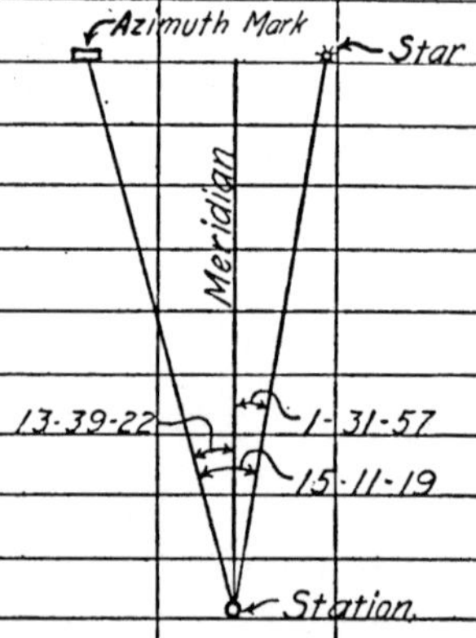

A.R.B., Observer.	Buff & Buff Transit No. 4849
A.H.H., Recorder.	Vernier: 00' 30"
	Sept. 20, 1915

Time of E Elong. of Polaris:

$\cos P = \frac{\tan \phi}{\tan \delta}$	$\log \tan \phi$	41-40	9.949 353
$t = 360° - P$	$\log \tan \delta$	88-51-19	1.699 374
$= 271°\text{-}01'\text{-}08''$	$\log \cos P$	88-58-52	8.249 979
$= 18^h\ 04^m\ 05^s$			

Sid T = RA + t = $1^h 30^m 18^s + 18^h 04^m 05^s = 19^h\ 34^m\ 23^s$

		h	m	s
A for Gh. M.N., Sept 20,	$11^h\ 53^m\ 11^s$			
Increase for λ ($6^h 06^m 06^s$),	1 00			
An,		11	54	11
Sid Int since M.N.,		7	40	12
Corr, Sid. to Solar,		−	1	15
LMT, 91°31'30"W, Astr., Sept. 20,		7	38	57
Std. T, Central, Sept 20, P.M.		7	45	03 ✓

From Table V:

Std. T, Central, of E Elong., Sept 20, P.M. 7 45.0 ✓

Azimuth of Polaris at Elong.:

$\sin Z = \frac{\cos \delta}{\cos \phi}$	$\log \cos \delta$	88-51-19	8.300 539
$Z = 1° 31' 57''$ ✓	$\log \cos \phi$	41-40	9.873 335
	$\log \sin Z$	1-31-57	8.427 204

From Table VII:

Az of Polaris at Elong., Sept. 20, 1° 32'.1 ✓

Azimuth by Polaris near West Elongation.
Station: Stk. near NE cor. of Washington and Madison Sts., Iowa City, Iowa.

$\phi = 41° 40'$, $\lambda = 91° 31' 30''$

Horiz. ∠s betw. Azimuth Mark and Star:

No. Rep.	Horiz. Circle Readings 'A' Vern.	'B' Vern.	Mean	Watch Readings
On Mk	00° 00' 00"	00' 00"	00' 00"	
1 D	21 16 00	16 00		7:01:00
2 D	—	—		7:04:30
1 R	—	—		7:08:40
2 R	85 02 30	03 00	02 45	7:11:20
	Arc passed over:		85° 02' 45"	
	Mean Angle:		21 15 41	

Watch Correction to Std. T.: $+00^m 30^s$

Std. T of Pointing	Interval 'T'	Correction 'C'
7:01:30	$18^m 36^s$	18"
7:05:00	22 06	25
7:09:10	26 16	36
7:11:50	28 56	43
		4)122
	Mean:	30".5

M.N.R., Observer.
A.R.C., Recorder

Buff & Buff Transit No. 4849.
Vernier: 00' 30."
April 3, 1916.

Time of W. Elong. of Polaris:

From Table V:

Std. T, Central, of W. Elong. Apr. 3, P.M. 6:42.9
or 6:42:54

Azimuth of Polaris at Elong:

From Table VII:

Az. at Elong., Apr. 3, 1° 31.7 or 1° 31' 42"

Bearing of Azimuth Mark:

Az. of Star at Elong:	1° 31' 42"
Correction:	− 30"
Az. of Star when sighted:	1 31 12, (Mean)
Horiz ∠, Star to Az. Mk.:	21-15-41
	1-31-12
	22 46 53

or, N 22° 46' 50" W, to nearest 10"

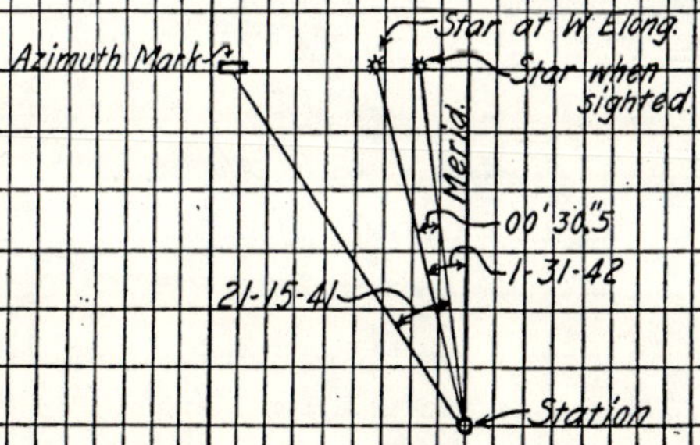

Azimuth by Polaris "at any Hour Angle."
Station: Stk. near NE cor. of Washington and Madison Sts., Iowa City, Iowa.

$\phi = 41° 40'$, $\lambda = 91° 31'30"$

No. Rep.	Horiz. Circle Readings 'A' Vern	'B' Vern.	Mean	Watch Readings	App. Altitude
On Mk.	00° 00'00"	00'00"	00'00"		
1 D	8 06 30	06 30		P.M. 10 29 00	
2 D				10 31 40	
3 D	24 22 30	22 30		10 33 32	
1 R				10 36 19	
2 R				10 38 07	
3 R	48 52 30	52 30	52 30	10 39 52	42° 49'
	Arc passed over:		48 52 30	6)63-28-30	
	Mean Angle:		8 08 45	10-34-45,	Mean
		Watch Corr.:		+ 00-44	
		Std. T:		10:35:29	

Bearing of Azimuth Mark:

8° 08'45"
0 02 32
8 11 17

or, N8° 11' 20" E, to nearest 10".

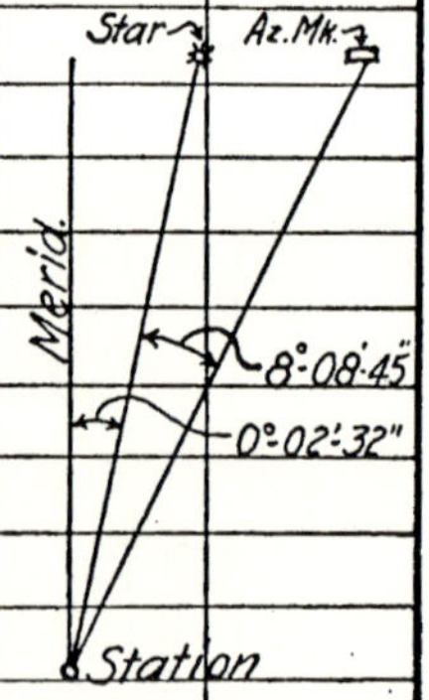

H.J.C., Observer. C.A.R., Recorder.	Buff & Buff Transit Vernier: 00'-30" Nov. 4, 1914

Computation of Hour Angle:

Std. T, Central,	Nov. 4, P.M.,		$10^h 35^m 29^s$
LMT, 91°31'30"W,	Nov. 4, Astr.,		10 29 23
Corr., Solar to Sid.,		+	1 43.4
Sid. Int. since M.N.			10 31 06.4
A at Gh. M.N., Nov. 4,	14-51-32.8		
Increase for λ (6-06-06),	1-00.1		
A_n			14 52 32.9
Sid. T,			25 23 39.3

$t = \text{Sid. T} - RA = 25^h 23^m 39^s.3 - 1^h 29^m 50^s.3$

$= 23^h 53^m 49^s.0 = 358° 27' 15"$

Computation of Azimuth:

$$\tan Z = \frac{\sin t}{\tan\delta \cos\phi - \sin\phi \cos t}$$

log tan δ,	88-51-16,	1.699 057
log cos φ	41-40	9.873 335
log tan δ cos φ,	37.359	1.572 392
log sin φ	41-40	9.822 688
log cos t	358-27-15	9.999 842
log sin φ cos t,	0.665	9.822 530
log Denom.	36.694	1.564 596
log sin t	358-27-15	8.430 983
log tan Z	0° 02' 32"	6.866 387

Azimuth by an Altitude of the Sun.
Station: Stk. near Eng. Building, Iowa City, Iowa.

$\phi = 41° 40'$, $\lambda = 91° 31' 30''$

Sun	No. Rep.	Horiz. Circle	Vert. Circle	Watch Rdg	
	On Mk.	00° 00'	—	—	
☉ (upper limb)	1D	279 10	43°27	$3^h 30^m 22^s$	
	2D	279 18	43 18	3 31 10	
☉ (lower limb)	1R	280 09	43 40	3 31 55	
	2R	280 19	43 32	3 32 43	
	On Mk.	00 00	—	—	
		4) 176	4) 117	4) 126 10	
	Mean:	279°44'00"	43°29'15"	$3^h 31^m 32^s$	
	Refraction Corr.:		− 1 01	−00 15	Watch Corr.
			43 28 14	$3^h 31^m 17^s$	Std. T.
	Parallax Corr.:		+ 00 07		
		h:	$43^h 28^m 21^s$		

Bearing of Azimuth Mark:

Horiz. Circle:		279°44'00"
∠ Z:		97 07 00
		376 51 00
		360 00 00
		16 51 00
	or,	N 16° 51' W,
		to nearest minute.

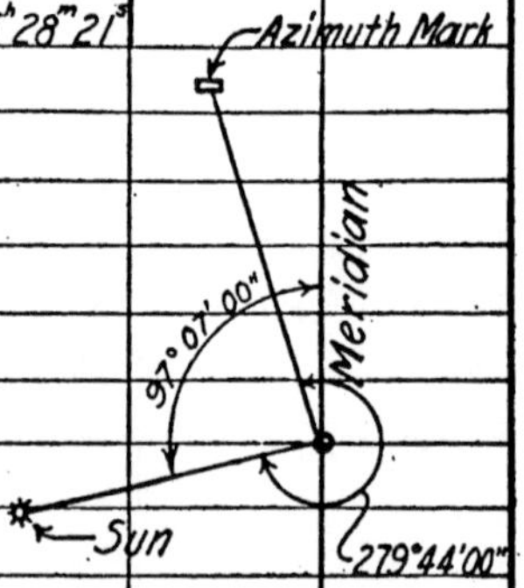

A.H.H., Observer. H.I.M., Recorder.	Gurley Transit. Verniers: 01'. June 9, 1916.

Declination of Sun:

δ at Gh. M.N., June 9,		+ 22° 56' 06"
λ of Std. Merid.:	$6^h 00^m 00^s$	
Std. T.,	3 31 17	
Since Gh. M.N.,	9 31 17	
Var. per hour	+ $12^s.57$	
Change in δ: (9.52 × $12^s.57$) :		+ 2 00
δ at mean of watch rdgs. :		+ 22 58 06

Azimuth of Sun:

$$\tan \tfrac{1}{2} Z = \sqrt{\frac{\cos(k+\phi)\ \cos(k+h)}{\sin k\ \cos(k+\delta)}}$$

ϕ,	41° 40'	log cos $(k+\phi)$	, 122° 36' 46"	, 9.731556
h,	43 28 21	log cos $(k+h)$	, 124 25 07	, 9.762229
δ,	+22 58 06	colog sin k	, 80 56 46	, 0.005445
	108 06 27	colog cos $(k+\delta)$	, 103 54 52	, 0.618934
	270 00 00			2 / 0.108164
2 /	161 53 33	log tan ½ Z	, 48 33 30	, 0.054082
k,	80° 56' 46"	Z	, 97° 07' 00"	

Azimuth by Equal Altitudes of β Cassiopeiae.
Station: Stk. near NE cor. of Washington and Madison Sts, Iowa City, Iowa.

Pointing	Apparent Altitude	Approx. Std. T of 1st Series	Approx. Std. T of 2nd Series
A	20° 00′	7:55 P.M.	3:10 A.M.
B	19 00	8:07	2:58
C	18 00	8:20	2:45

Angle	Single	Double	∠ Double ÷ 2	∠/2	
A_1-⊙-A_2	53°-30′-00″	106°-59′-30″	53°-29′-45″	26°-44′-52″	= A_1-⊙-M_A
B_1-⊙-B_2	50-54-00	101-47-30	50-53-45	25-26-52	= B_1-⊙-M_B
C_1-⊙-C_2	48-05-30	96-11-00	48-05-30	24-02-45	= C_1-⊙-M_C

Stake lined in and centered at M, mean of directions of bisectors, M_A, M_B, M_C.

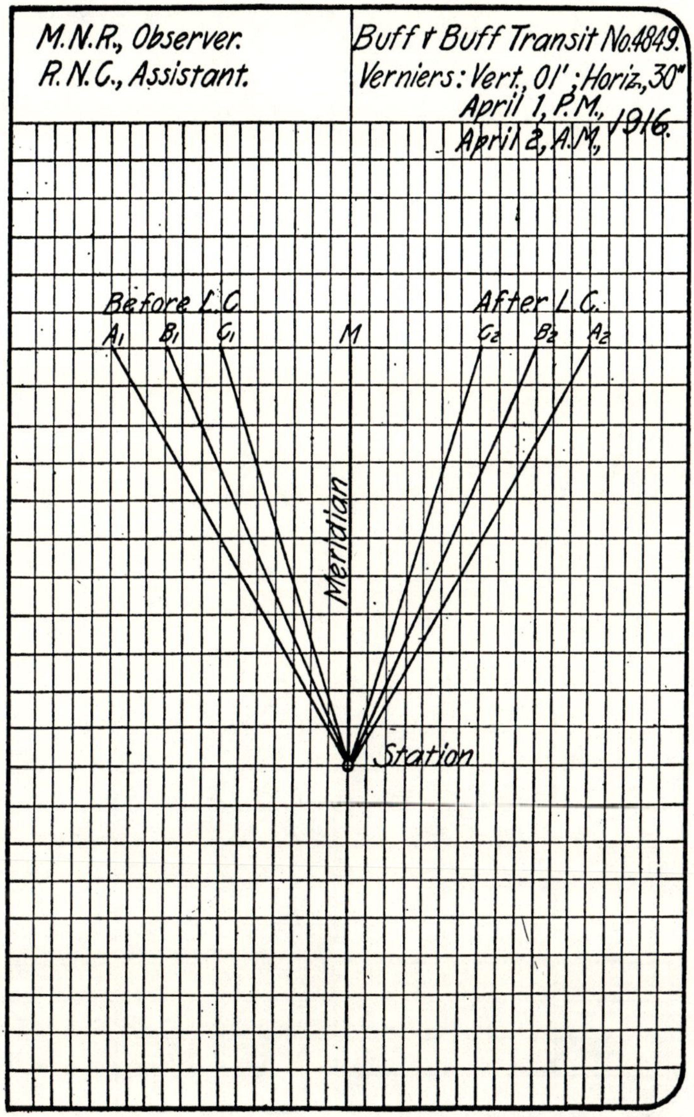
M.N.R., Observer.
R.N.C., Assistant.
Buff & Buff Transit No.4849.
Verniers: Vert., 01'; Horiz., 30"
April 1, P.M.,
April 2, A.M., 1916.
Before L.C.
After L.C.
A1
B1
C1
M
C2
B2
A2
Meridian
Station

Time by Transits of Stars.
Station: Stk. near NE cor. of Washington and Madison Sts, Iowa City, Iowa.
$\phi = 41° 40'$, $\lambda = 91° 31' 30''$

Star	Mag.	Approx. Merid. Alt.	Comp. LMT of Trans.	Watch Rdg. at Trans.	Error of Watch on LT
γ Hydrae	3.3	25° 36'	$8^h 25^m 35^s$	$8^h 31^m 10^s$	$+5^m 35^s$
α Virginis	1.2	37 36	8 31 59	8 37 30	5 31
ξ Virginis	3.4	48 10	8 41 36	8 46 47	5 11
τ Virginis	4.3	50 17	9 08 29	9 14 10	5 41
π Hydrae	3.5	22 03	9 12 42	9 18 02	5 20
κ Virginis	4.3	38 27	9 19 30	9 25 05	+5 35
					6/173
			Mean Error:		$+5^m 29^s$

Watch Correction to LMT: $-5^m 29^s$.
" " " StdT: $+0^m 37^s$

First three observations made with telescope direct, last three with telescope inverted.

M. N. R., Observer.
R. N. C., Recorder.

Bausch & Lomb Transit.

June 3, 1916

Limting Altitudes to be 10° and 50°, $\phi = 41°40'$;
∴ Limiting declinations (for southern stars) are: $+1°40'$ and $-38°20'$.

For limiting right ascensions for work between 8:30 and 9:30 P.M.; change 8:30 P.M., Std. T, to Sid. T.

Std. T, Central, June 3, P.M.,		8^h	30^m	00^s
LMT, 91°31'30" W, June 3, Astr.,		8	23	54
Corr., Solar to Sid.,	+		1	23
Sid. Int. since M.N.		8	25	17
A at Gh. M.N., June 3, $4^h 46^m 26^s$				
Increase for λ (6.06.06), 1 00				
A_n,		4	47	26
Sid. T,		13	12	43

$$RA = \text{Sid. } T - t$$
$$= 13^h 12^m 43^s - 0^h = 13^h 12^m 43^s$$

∴ Limiting right ascensions are: $13^h 13^m$, and $14^h 13^m$.

Time by Transit of the Sun.
Station: Stk. near NE cor. of Washington and Madison Sts., Iowa City, Iowa.

$\phi = 41°40'$, $\lambda = 91°31'30''$

Sun	Watch Reading at Transit
d	$12^h\ 02^m\ 35^s$
p	12 04 45
	2) 7 20
Mean:	12 03 40

Watch Corr. to LMT: $-\ 5^m\ 43^s$
" " " Std. T: $+\ 0^m\ 23^s$

M.N.R., Observer.
A.H.H., Recorder.

Gurley Transit.

June 3, 1916.

Time of Transit of Sun's Center:

L A T, 91° 31' 30" W, June 3,		0^h	00^m	00^s
λ W of Gh.,	6 06 06			
Since Gh. App. N, Jun3,	6 06 06			
E (from Naut Al as for Mean Time)		−	02	03
L M T, 91° 31' 30" W, June 2, Astr.,		23	57	57
L M T, do. , June 3, Civil, A.M.,		11	57	57
Std. T, Central, June 3, P.M.,		12	04	03

INDEX

Spotted something that isn't quite right?

Let us know and we'll fix it for future reprints.
And as a thank you for your diligence, we'll
send you a free Sothis book of your choice.

publisher@sothis.press

This page intentionally left blank

Made in United States
North Haven, CT
05 November 2026

21817318R00085